Communications
in Computer and Information Science 64

Dominik Ślęzak Tai-hoon Kim
Yanchun Zhang Jianhua Ma
Kyo-il Chung (Eds.)

Database
Theory and Application

International Conference, DTA 2009
Held as Part of the Future Generation
Information Technology Conference, FGIT 2009
Jeju Island, Korea, December 10-12, 2009
Proceedings

 Springer

Volume Editors

Dominik Ślęzak
University of Warsaw and Infobright Inc., Poland
E-mail: slezak@infobright.com

Tai-hoon Kim
Hannam University, Daejeon, South Korea
E-mail: taihoonn@hnu.kr

Yanchun Zhang
Utrecht University, The Netherlands
E-mail: y.zhang@geo.uu.nl

Jianhua Ma
Hosei University, Tokyo, Japan
E-mail: jianhua@hosei.ac.jp

Kyo-il Chung
ETRI, South Korea
E-mail: kyoil@etri.re.kr

Library of Congress Control Number: 2009940117

CR Subject Classification (1998): H.2, K.8.1, H.2.8, H.3.5, I.2.6, H.5, I.7

ISSN   1865-0929
ISBN-10   3-642-10582-3 Springer Berlin Heidelberg New York
ISBN-13   978-3-642-10582-1 Springer Berlin Heidelberg New York

springer.com

© Springer-Verlag Berlin Heidelberg 2009
Printed in Germany

Typesetting: Camera-ready by author, data conversion by Scientific Publishing Services, Chennai, India
Printed on acid-free paper   SPIN: 12807369   06/3180   5 4 3 2 1 0

Foreword

As future generation information technology (FGIT) becomes specialized and fragmented, it is easy to lose sight that many topics in FGIT have common threads and, because of this, advances in one discipline may be transmitted to others. Presentation of recent results obtained in different disciplines encourages this interchange for the advancement of FGIT as a whole. Of particular interest are hybrid solutions that combine ideas taken from multiple disciplines in order to achieve something more significant than the sum of the individual parts. Through such hybrid philosophy, a new principle can be discovered, which has the propensity to propagate throughout multifaceted disciplines.

FGIT 2009 was the first mega-conference that attempted to follow the above idea of hybridization in FGIT in a form of multiple events related to particular disciplines of IT, conducted by separate scientific committees, but coordinated in order to expose the most important contributions. It included the following international conferences: Advanced Software Engineering and Its Applications (ASEA), Bio-Science and Bio-Technology (BSBT), Control and Automation (CA), Database Theory and Application (DTA), Disaster Recovery and Business Continuity (DRBC; published independently), Future Generation Communication and Networking (FGCN) that was combined with Advanced Communication and Networking (ACN), Grid and Distributed Computing (GDC), Multimedia, Computer Graphics and Broadcasting (MulGraB), Security Technology (SecTech), Signal Processing, Image Processing and Pattern Recognition (SIP), and u- and e-Service, Science and Technology (UNESST).

We acknowledge the great effort of all the Chairs and the members of advisory boards and Program Committees of the above-listed events, who selected 28% of over 1,050 submissions, following a rigorous peer-review process. Special thanks go to the following organizations supporting FGIT 2009: ECSIS, Korean Institute of Information Technology, Australian Computer Society, SERSC, Springer LNCS/CCIS, COEIA, ICC Jeju, ISEP/IPP, GECAD, PoDIT, Business Community Partnership, Brno University of Technology, KISA, K-NBTC and National Taipei University of Education.

We are very grateful to the following speakers who accepted our invitation and helped to meet the objectives of FGIT 2009: Ruay-Shiung Chang (National Dong Hwa University, Taiwan), Jack Dongarra (University of Tennessee, USA), Xiaohua (Tony) Hu (Drexel University, USA), Irwin King (Chinese University of Hong Kong, Hong Kong), Carlos Ramos (Polytechnic of Porto, Portugal), Timothy K. Shih (Asia University, Taiwan), Peter M.A. Sloot (University of Amsterdam, The Netherlands), Kyu-Young Whang (KAIST, South Korea), and Stephen S. Yau (Arizona State University, USA).

We would also like to thank Rosslin John Robles, Maricel O. Balitanas, Farkhod Alisherov Alisherovish, and Feruza Sattarova Yusfovna – graduate students of Hannam University who helped in editing the FGIT 2009 material with a great passion.

October 2009

Young-hoon Lee
Tai-hoon Kim
Wai-chi Fang
Dominik Ślęzak

Preface

We would like to welcome you to the proceedings of the 2009 International Conference on Database Theory and Application (DTA 2009), which was organized as part of the 2009 International Mega-Conference on Future Generation Information Technology (FGIT 2009), held during December 10–12, 2009, at the International Convention Center Jeju, Jeju Island, South Korea.

DTA 2009 focused on various aspects of advances in data processing, databases and data warehouses, in conjunction with computational sciences, mathematics and IT. It provided a chance for academic and industry professionals to discuss recent progress in the related areas. We expect that the conference and its publications will be a trigger for further related research and technology improvements in this important subject.

We would like to acknowledge the great effort of all the Chairs and members of the Program Committee. Out of 80 submissions to DTA 2009, we accepted 23 papers to be included in the proceedings and presented during the conference. This gives an acceptance ratio firmly below 30%. Two of the papers accepted to DTA 2009 were published in the special FGIT 2009 volume, LNCS 5899, by Springer. The remaining 21 accepted papers can be found in this CCIS volume.

We would like to express our gratitude to all of the authors of submitted papers and to all of the attendees, for their contributions and participation. We believe in the need for continuing this undertaking in the future.

Once more, we would like to thank all the organizations and individuals who supported FGIT 2009 as a whole and, in particular, helped in the success of DTA 2009.

October 2009

Dominik Ślęzak
Tai-hoon Kim
Yanchun Zhang
Jianhua Ma
Kyo-il Chung

Organization

Organizing Committee

General Chairs	Yanchun Zhang (Victoria University, Australia)
	Dominik Ślęzak (University of Warsaw and Infobright, Poland)
Program Chairs	Jianhua Ma (Hosei University, Japan)
	Kyo-il Chung (ETRI, Korea)
Steering Chairs	Tai-hoon Kim (Hannam University, Korea)
	Wai-chi Fang (National Chiao Tung University, Taiwan)

Program Committee

Alfredo Cuzzocrea
Anne James
Aoying Zhou
Chunsheng Yang
Damiani Ernesto
Daoqiang Zhang
David Taniar
Djamel Abdelakder
Emiran Curtmola
Feipei Lai
Fionn Murtagh
Gang Li
Guoyin Wang
Haixun Wang
Hans-Joachim Klein
Hiroyuki Kawano
Hiroshi Saka
Hui Xiong
Hui Yang
Janusz Borkowski
Jason T.L. Wang
Jesse Z. Fang
Jian Lu
Jian Yin
Joel Quinqueton
Joshua Z. Huang

Jun Hong
Junbin Gao
Karen Renaud
Kay Chen Tan
Kenji Satou
Keun Ho Ryu
Krzysztof Stencel
Lachlan McKinnon
Ladjel Bellatreche
Laura Rusu
Li Ma
Longbing Cao
Lucian N. Vintan
Mark Roantree
Masayoshi Aritsugi
Miyuki Nakano
Ozgur Ulusoy
Pabitra Mitra
Pang-Ning Tan
Peter Baumann
Richi Nayak
Rosa Meo
Sanghyun Park
Sang-Wook Kim
Sanjay Jain
Shu-Ching Chen

Shyam Kumar Gupta
Stephane Bressan
Tadashi Nomoto
Tao Li Florida
Tetsuya Yoshida
Theo Härder
Tomoyuki Uchida
Toshiro Minami
Vasco Amaral
Veselka Boeva
Victoria Eastwood
Vicenc Torra
Wei Wang
Weining Qian
Weiyi Meng
William Zhu
Xiaohua (Tony) Hu
Xiao-Lin Li
Xuemin Lin
Yan Wang
Yang Yu
Yang-Sae Moon
Yiyu Yao
Young-Koo Lee
Zhuoming Xu

Table of Contents

Steganalysis for Reversible Data Hiding

Ho Thi Huong Thom[1], Ho Van Canh[2], and Trinh Nhat Tien[3]

[1] Faculty of Information Technology, Hai Phong Private University, Vietnam
thomhth@hpu.edu.vn
[2] Dept. of Professional Technique, Ministry of Public Security, Vietnam
hovancanh@gmail.com
[3] College of Technology, Vietnam National University, HaNoi, Vietnam
tientn@vnu.edu.vn

Abstract. In recent years, several lossless data hiding techniques have been proposed for images. Lossless data embedding can take place in the spatial domain or in the transform domain. They utilized characteristics of the difference image or the transform coefficient histogram and modify these values slightly to embed the data. However, after embedding message bits these steganography changed the nature of the difference image histogram or the transform coefficient histogram gradually. In this paper, we introduce two new steganalytic techniques based on the difference image histogram and the transform coefficient histogram. The algorithm can not only detect existence of secret messages in images which are embedded by above methods reliably, but also estimate the amount of hidden messages exactly.

Keywords: Steganography, Steganalysis, Cover Image, Stego Image, The Difference Image, Histogram Shifting, Lossless Data Hiding, Integer Wavelets.

1 Introduction

Steganography is the art of secret communication, its purpose is to convey messages secretly by concealing the very existence of the message. Similar to cryptanalysis, steganalysis attempts to defeat the goal of steganography. It is the art of detecting the existence of hidden information. Digital images, videos, sound files and other computer files that contain perceptually irrelevant or redundant information can be used as "cover" or carriers to hide secret messages. After embedding a secret message into the cover-image, a so-called stego-image is obtained.

In recent years, several lossless data hiding techniques have been proposed for images. Lossless data embedding can take place in the spatial domain [1, 2, 3], or in the transform domain [4, 5]. Lee et al [3] proposed a lossless data embedding technique (we assume that the technique name DIH method), which utilizes characteristics of the difference image and modifies pixel values slightly to embed the secret data. Xuan et al [5] proposed a histogram shifting method in integer wavelet transform domain (we assume that the technique name IWH method). This algorithm hides message into integer wavelet coefficients of high frequency subbands.

In this paper, we propose two new steganalytic methods based on the difference image histogram and integer wavelets transform. Former can detect stego images

D. Ślęzak et al. (Eds.): DTA 2009, CCIS 64, pp. 1–8, 2009.

using Lee's steganography (DIH method) and latter detects stego images using Xuan's method (IWH method). The algorithms can also estimate the embedded data length reliably. In the next section, we describe again the details of Lee's and Xuan's steganography. In section 3, we present our proposed steganalytic methods. Our experimental results are given in section 4. Finally, concluding in section 5.

2 Reversible Data Hiding

2.1 Lossless Data Hiding Based on Histogram Modification of Difference Image

In this subsection, we describe again details of Lee and his colleague's lossless data hiding method using the histogram modification of the difference image [3].

2.1.1 Watermark Embedding

They assume that embedded data is a binary Logo sequence B(m,n) of size PxQ pixels, they combine a binary random sequence generated by the user key A(l) of length PxQ bits with B(m,n) using the bit – wise XOR operation, they get a binary watermark sequence W(m,n) of size PxQ.

For a grayscale image I(i, j) of size MxN pixels, they form the difference image D(i, j) of size MxN/2 from the original image. D(i, j) = I(i, 2j+1)-I(i,2j), $0 \le i \le M-1$, $0 \le j \le \frac{N}{2}-1$, where I(i, 2j+1), I(i,2j) are the odd-line field and the even-line field, respectively.

For watermark embedding, they empty the histogram bins of -2 and 2 by shifting some pixel values in the difference image. If the difference value is greater than or equal to 2, they add one to the odd-line pixel. If the difference value is less than or equal to -2, they subtract one from the odd-line pixel. Then, the difference image is modified $\tilde{D}(i,j) = \tilde{I}$ (i, 2i+1) - $\tilde{I}$ (i, 2j) where $\tilde{I}$ (i, 2j+1) and $\tilde{I}$ (i,2j) are the odd-line field and the even-line field of the modified image, respectively.

In the histogram modification process, the watermark W(m,n) is embedded based on the modified difference image $\tilde{D}(i,j)$. The modified difference image is scanned. Once a pixel with the difference value of -1 or 1 is encountered, they check the watermark to be embedded. If the bit to be embedded is 1, they move the difference value of -1 to -2 by subtracting one from the odd-line pixel or 1 to 2 by adding one to the odd-line pixel. If the bit to be embedded is 0, they skip the pixel of the difference image until a pixel with the difference value -1 or 1 is encountered. In this case, there is no change in the histogram. Therefore, the watermarked fields I_w(i, 2j+1) and I_w(i,2j) are obtained by

$$I_w = \begin{cases} \tilde{I}(i,2j+1)+1 & \text{if } \tilde{D}(i,j)=1 \text{ and } W(m,n)=1 \\ \tilde{I}(i,2j+1)-1 & \text{if } \tilde{D}(i,j)=-1 \text{ and } W(m,n)=1 \\ \tilde{I}(i,2j+1) & \text{otherwise} \end{cases} \qquad (1)$$

and

$$I_w(i,2j)=I(i,2j). \qquad (2)$$

2.1.2 Watermark Extraction and Recovery

Calculating the difference image $D_e(i,j)$ from the received watermarked image $I_e(i,j)$. The whole difference image is scanned. If the pixel with the difference value of -1 or 1 is encountered, the bit 0 is retrieved. If the pixel with the difference value of -2 or 2 is encountered, the bit 1 is retrieved. In this way, the embedded watermark $W_e(m,n)$ can be extracted.

Finally, we reverse the watermarked image back to the original image by shifting some pixel values in the difference image. The whole difference image is scanned once again. If the difference value is less than or equal to -2, they add one to the odd-line pixel. If the difference value is greater than or equal to 2, they subtract one from the odd-line pixel.

2.1.3 Lossless Image Recovery

The proposed scheme cannot be completely reversed because the loss of information occurs during addition and subtraction at the boundaries of the grayscale range (at the gray level 0 and 255). In order to prevent this problem, they adopt modulo arithmetic for watermark addition and subtraction. For the odd –line field $I(i,2j+1)$, they define the addition modulo c as

$$I(i,2j+1) +_c 1 = (I(i,2j+1)+1) \bmod c \tag{3}$$

where c is the cycle length. The substraction modulo c is defined as

$$I(i,2j+1) -_c 1 = (I(i,2j+1)-1) \bmod c \tag{4}$$

The reversibility problem arises from pixel that is truncated due to overflow or underflow. Therefore, they use $+_c$ and $-_c$ instead of + and – only when truncation due to the occurrence of overflow or underflow. In other words, they have only to consider $255 +_c 1$ and $0 -_c 1$.

In the receiving side, it is necessary to distinguish between the case when, for example, $I_e(i, 2j+1)=255$ was obtained as $I(i,2j+1)+1$ and $I(i,2j+1) -_{256} 1$. They assume that no abrupt change between two adjacent pixels occurs. If there is a significant difference between $I_e(i, 2j+1)$ and $I_e(i, 2j)$, we estimate that $I(i,2j+1)$ was manipulated by modulo arithmetic.

$$\begin{cases} I(i,2j+1)+1 & \text{if } | I_e(i,2j+1) - I_e(i,2j) | \le \tau \\ I(i,2j+1) -_{256} 1 & \text{otherwise} \end{cases} \tag{5}$$

Where τ is a threshold value. Similarly, $I_e(i,2j+1)=0$ is estimated as

$$\begin{cases} I(i,2j+1)-1 & \text{if } | I_e(i,2j+1) - I_e(i,2j) | \le \tau \\ I(i,2j+1) +_{256} 1 & \text{otherwise} \end{cases} \tag{6}$$

2.2 Lossless Data Hiding Based on Integer Wavelet Histogram Shifting

In the subsection, we describe again the detail of IWH method. Since it is required to reconstruct the original image with no distortion, Xuan et al [5] use the integer lifting scheme wavelet transform. After integer wavelet transform, it has four sub-bands.

They will embed the information into three high frequency sub-bands. IWH method is presented as follows.

2.2.1 Data Embedding Algorithm

Assume there are M bits which are supposed to be embedded into a high frequency subband of IWT. We embed the data in the following steps:

(1) Let a threshold T>0 be the number of the high frequency wavelet coefficients in [-T, T] is greater than M. And set the Peak=T.

(2) In the wavelet histogram, move the histogram (the value is greater than Peak) to the right-hand side by one unit to leave a zero-point at the value Peak+1. Then embed data in this point. Scanning all of IWT coefficients in the high frequency subband. Once an IWT coefficient of value "Peak" is encountered, if the to be embedded bit is 1, this coefficient's value will be added by 1, i.e, becoming "Peak+1". If the to be embedded bit is 0, the coefficient's value remain to be "Peak".

(3) If there are to-be-embedded data remaining, let Peak = (-Peak), and move the histogram (less than Peak) to the left-hand side by 1 unit to leave a zero-point at the value (-Peak-1). And embed data in this point.

(4) If all the data are embedded, then stop here and record the Peak value as stop peak value, S. Otherwise, Peak =(-Peak-1), go back to (2) to continue to embed the remaining to-be-embedded data.

2.2.2 Data Extraction Algorithm

Data extraction is the reverse process of data embedding. Assume the stop peak value is S, the threshold is T. The data extraction process is performed as follows:

(1) Set Peak = S.

(2) Decode with the stop value Peak. (In what follows, assume Peak>0). When an IWT coefficient of value "Peak+1" is met, bit "1" is extracted and the coefficient's value reduces to "Peak". When the coefficient of value "Peak" is met, bit "0" is extracted. Extract all the data until Peak+1 becomes a zero-point. Move all the histogram (greater than Peak+1) to the left-hand by one unit to cover the zero-point.

(3) If the extracted data is less than M, set Peak= -Peak. Continue to extract data until it becomes a zero-point in the position (Peak-1). Then move histogram (less than Peak-1) to the right-hand side by one unit to cover the zero-pint.

(4) If all the hidden bits have been extracted, stop. Otherwise, set Peak=-Peak+1, go back to (2) to continue to extract the data.

3 Proposed Steganalytic Methods

3.1 Steganalytic Method for DIH Method

After embedding a set of message into a set of original image using DIH method get a set of stego image, we form the difference image D(i,j) and calculate the histogram of

 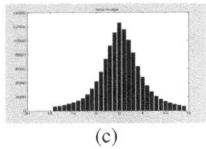 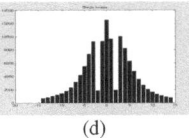

a)   b)   (c)   (d)

Fig. 1. Test images: a) Lena original image, b) Binary logo image. (c) The difference image histogram of original image, (d) The difference image histogram of stego image.

D(i,j) for each image in the original image set and the stego image set, we found out that DIH method changed natural of the difference image histogram of typical image significantly as the example in [3].

We use a original Lena grayscale image of size 512 x 512 pixels (Fig.1. (a)), we perform the difference image D(i,j) and calculate histogram of all D(i,j) that is shown in Fig.1. (c). We then embed a watermark which is a binary logo image of size 128x56 pixels, equivalent to a binary sequence of 7,168 bits (following the example of [3], see Fig.1. (b)) into Lena image using DIH. We get the difference image histogram in Fig.1. (d).

From Fig.1 (c) and (d), comparing difference between the original histogram and the histogram after data embedding we see easily that DIH changed the difference value of -2 and 2 considerably. In any typical image, the histogram value of the difference value -2 and 2 (denote h_2 and h_{-2}) is always greater than the histogram value of the difference value -3 and 3 (denote h_3 and h_{-3}), it mean that sum of h_2 and h_{-2} greater than sum of h_3 and h_{-3} τ time ($\tau \geq 1$ is a threshold). In other words, in a stego image, the sum of h_2 and h_{-2} is less than of h_3 and h_{-3} τ time.

In certain cases, the sum of h_2 and h_{-2} is less than of h_3 and h_{-3} τ time, we can conclude that this image is a stego image using DIH method and we can also estimate data length which was embedded in original image as the following analysis.

We assume that the to be embedded data is a binary sequence W of n bits with the number of bit "1" equals to the number of bit "0", approximately. After embedding the hidden sequence into an original image, DIH method will shift a part n/2 of h_1 and h_{-1} to h_2 and h_{-2} to store n/2 message bits, it means a part n/2 of the original h_1 and h_{-1} now becomes n/2 of h_2 and h_{-2}, remaining n/2 data bits are embedded into a part n/2 of h_1 and h_{-1}. Therefore, the hidden data length equals to $(h_2+h_{-2})*2$, approximately. In our experimentation, we get high reliable result with $\tau = 1.15$.

Applying to above example, we estimate the embedded data length L=7034 bits which are embedded in Lena image.

In special case, users doesn't use different value of -2 and 2 to embed data, they choose other different values. We change the method a little. The difference image histogram is scanned. Once a pair of histogram of i+1 and − (i+1) is greater than of i and −i with τ time, we set i be the location to estimate the embedded data length.

3.2 Steganalytic Method for IWH Method

To estimate message length in stego image using IWH method, we first give analysis of occurrences in watermarking process as the three following experiments:

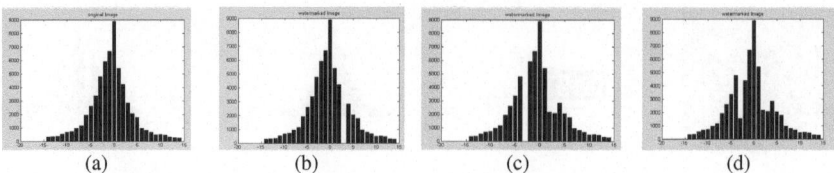

(a)   (b)   (c)   (d)

Fig. 2. An example showing how a zero point is generated and payload data embedding process: (a) original histogram, (b) histogram after a zero point is created, (c) histogram after data embedding at Peak =2 and then a new zero point is created at new next Peak, (d) histogram after data remaining embedding with new Peak.

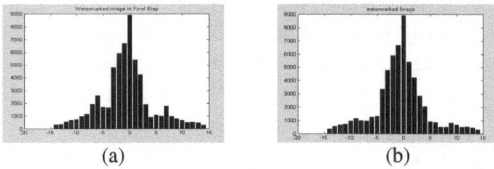

(a)   (b)

Fig. 3. Another example showing how payload data embedding process: (a) the histogram after data embedding with chosen T=4, (b) histogram after data embedding with chosen T=6.

In the first experiment, we use also Lena image and Logo image in section 3.1 to test. After integer wavelet transform, we calculate the histograms of high frequency subbands (see Fig. 2. (a)). We next embed payload data (that is the binary sequence from Logo image) in to the high frequency subbands with T=2 using IWH method. We get S=-2 and calculate again the high frequency subbands that is shown in Fig. 2. (d). The data embedding process performs via some steps: the first and second step embeds data in the point 2 and 3 (see Fig.2. (b), (c)) but there are to be embedded data remaining, the process performs the third and fourth step with T=-2 to embed data (see Fig.2. (c), (d)). In the second experiment, we use also the Lena original image and Logo watermark with T=4, we then get S=3. In this case, the histogram is changed much that is shown in Fig.3. (a). In the third experiment, we use the same input with T=6, we then get S=-5. In this case, the histogram is changed clearly that is shown in Fig.3. (b).

We compare difference between the histogram of typical image and of stego image, we found out that, in typical image, $h_0 > h_1 > h_2 > h_3 > \ldots$ and $h_0 > h_{-1} > h_{-2} > h_{-3} > \ldots$ where h_i is histogram value of integer wavelet coefficient i.

The stego image in the first experiment, we get $h_4 > h_3$, $h_3 \approx h_2$, $h_{-4} > h_{-3}$, $h_{-3} < h_{-2}$.

The stego image in the second experiment, we get $h_5 \approx h_6$, $h_{-5} \approx h_{-4}$, $h_4 < h_3$, $h_4 < h_5$

The stego image in the third experiment, we get $h_7 \approx h_8$, $h_5 \approx h_6$, $h_{-7} \approx h_{-8}$, $h_{-5} \approx h_{-6}$.

We explain detail of these problems of the third experiment. IWT method first shift the part of histogram with value greater than T=6 towards the right hand side by one unit. After data embedding we get $h_6 \approx h_7$. Due to data remaining, Peak T=6 change new T=-6, at this step, amount of remaining data fit histogram value -6, so after data embedding, $h_{-6} \approx h_{-7}$. Next, T=-6 becomes T=5, h_6 and h_7 move to h_7 and h_8, and $h_5 \approx h_6$ due to remaining available payload. T =5 change to T=-5 again, h_{-6} and h_{-7} move to

h_{-7} and h_{-8} to embed data, remaining payload embed into a part of h_{-5}, it makes a part of h_{-5} become h_{-6} (due to a number of remaining data equal h_{-5}, so $h_{-5} \approx h_{-6}$). Finally, IWT method finish and set S=T=-5.

From above analyses, we give generally steganalytic algorithm estimating length of payload data as follows:

(1) Initiate data length L=0, scan all histogram of value i (i>=0, i <=max (all integer wavelet coefficient of high subbands)), if the first $(h_i + h_{i+1})/2 < h_{i+2}$ is met, stop scanning, let Peak = i be first location to estimate data length.

(2) if $h_{Peak} \approx h_{Peak+1}$, $L=L+h_{Peak}+h_{peak+1}$; set Peak = -Peak and perform next step 3. Otherwise, perform step 4.

(3) if $h_{Peak} \approx h_{Peak+1}$, $L=L+h_{Peak}+h_{peak+1}$; set Peak = -Peak - 1 and return step 2. Otherwise, perform step 4.

(4) if $h_{Peak+1}<h_{Peak+2}$ and $h_{Peak+1}< h_{Peak}$ then $L= L+2*h_{Peak+1}$. The process finishes here.

Applying the algorithm to three above experiments, we estimate the embedded data lengths that are shown in table 1.

Table 1. Experimental result on Lena image

Embedded data length	Chosen threshold T	Gotten Stop value S	Estimated data length
7168	2	-2	7231
7168	4	3	6998
7168	6	-5	7177

4   Experimental Results

We have a set of images, it includes 600 grayscale images with 350 standard test JPEG image size of 768x512 or 512x512 pixels they were downloaded from [6], [7], and 250 test JPEG images with 1280x960 pixels they were created from my digital camera, all images are then converted to grayscale images by Photoshop CS2 software.

From the above set, we create two new sets. First set includes 1200 images with 600 original images and 600 stego-images which are embedded the same secret binary sequence of 6000 bits into corresponding 600 original images by DIH method. Second set includes also 1200 images with 600 original images and 600 stego-images which are embedded the same secret binary sequence of 6000 bits into corresponding 600 original images by IWH method. Then, we use our two proposed steganalytic methods to detect cover image and stego image and estimate embedded data length for two sets, respectively. The test results are shown in Fig. 4. (a) and (b). There the horizontal axis represents image number # and the vertical axis represents the embedded data length corresponding image number #.

In the first experimental result, our first steganalytic method detected exactly 600 original images (100 %) and the mean of estimated data lengths equals to 6397.5. In the second experimental result, our second steganalytic method detects exactly 544 original images (90.67 %) and the mean of estimated data lengths equals to 5356.5.

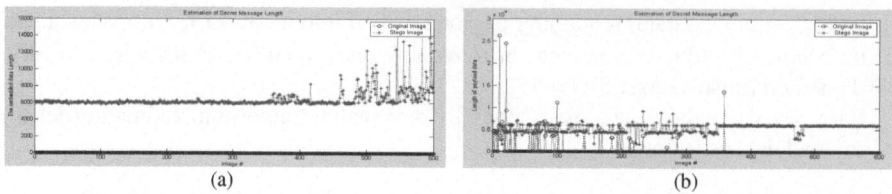

Fig. 4. Experimental results: (a) Estimated message length for the database using DIH method, (b) Estimated message length for the database using IWH method

Comparing accuracy between two methods we found that error estimation concentrates on very noisy images. So we give several factors that influence the accuracy of the estimation: **Noise:** For very noisy images, it makes histogram value of difference image and high integer wavelet coefficient become closer. So our steganalytic methods can detect falsely. **Chosen Peak:** If T is chosen be greater than 10 for IWH method, our method 2 will be very hard to estimate embedded data length in stego image.

5 Conclusions

This paper proposes two new steganalytic algorithms, which bases on histogram of the difference image and integer wavelet coefficient of high subbands. Experimental results show that our methods are reliable. However, it is hard to detect stego-image with two factors which are shown in section 4. We acknowledge that there are many elements in our algorithms that can be changed or replaced with other elements.

References

1. Honsinger, C., Jone, P., Rabbani, M., Stoffel, J.: Lossless recovery of an original image containing embedded data. US Patent: 6,278,791 B1 (2001)
2. Ni, Z., Shi, Y., Ansari, N., Su, W.: Reversible data hiding. In: Proc. ISCAS, pp. 912–915 (2003)
3. Lee, S.-K., Suh, Y.-H., Ho, Y.-S.: Lossless Data Hiding Based on Histogram Modification of Difference Images. In: Aizawa, K., Nakamura, Y., Satoh, S. (eds.) PCM 2004. LNCS, vol. 3333, pp. 340–347. Springer, Heidelberg (2004)
4. Xuan, G., Zhu, J., Chen, J., Shi, Y., Ni, Z., Su, W.: Distortionless data hiding based on integer wavelet transform. IEEE Electrionics Letters, 1646–1648 (2002)
5. Xuan, G., Yao, Q., Yang, C., Gao, J., Chai, P., Shi, Y.Q., Ni, Z.: Lossless Data Hiding Using Histogram Shifting Method Based on Integer Wavelets. In: Shi, Y.Q., Jeon, B. (eds.) IWDW 2006. LNCS, vol. 4283, pp. 323–332. Springer, Heidelberg (2006)
6. CBIR image database, University of Washington,
   http://www.cs.washington.edu/research/imagedatabase/groundtruth/
7. USC-SIPI Image Database,
   http://sipi.usc.edu/services/database/Database.html

An Incremental View Maintenance Approach Using Version Store in Warehousing Environment

AbdulAziz S. Almazyad, Mohammad Khubeb Siddiqui, Yasir Ahmad, and
Zafar Iqbal Khan

College of Computer Engineering and Sciences,
King Saud University, KSA
{mazyad,khubeb,yasirahmad,zkhan1}@ksu.edu.sa

Abstract. Data warehouse is a repository of information collected from multiple, possibly heterogeneous, autonomous and distributed databases. The information stored in the warehouse is in the form of Materialized views. In data warehouses materialized views are used to pre-compute and store aggregated data such as sums and averages. Materialized views are the derived relation, which are stored as the relation in the database, when some update occurs in the parent relation all its child relations also gets updated by viewing to maintain the consistency and convergence [3] of the database. In this paper we present a new method of incremental view maintenance with the inclusion of some existing approaches [2][3]. We utilized the concept of version store [2] for older versions of tables that have been updated at the source.

Keywords: Data Warehouse, Materialized View, Version Store, View Manager, View Maintenance.

1 Introduction

Data warehouse means storage of data (may be in the size of terabytes of disk storage), data warehouse is a copy of transaction data specifically structured for querying and reporting, which stores volume of historical data.

A data warehouse can be normalized or de normalized. It can be a relational database, multidimensional database, hierarchical database, object database, etc. It should be: Subject-oriented, Integrated, Non-volatile, Time-Variant, Accessible, and Process-Oriented. In data warehouses, materialized view act as a cache, a copy of data that can be quickly accessed because indexes are built up over that, you can use materialized views to pre compute and store aggregated data. Materialized views in these environments are often referred to as summaries, because they store summarized data. They can also be used to pre compute joins with or without aggregations. A materialized view eliminates the overhead associated with expensive joins and aggregations for a large or important class of queries. Materialized views are of three types Materialized Views with Aggregates, Materialized Views Containing Only Joins and Nested Materialized Views. Maintaining a view is of the most important task in warehousing environment.

The process of updating a materialized view in response to change to the underlying data is called view maintenance [1].

D. Ślęzak et al. (Eds.): DTA 2009, CCIS 64, pp. 9–16, 2009.
© Springer-Verlag Berlin Heidelberg 2009

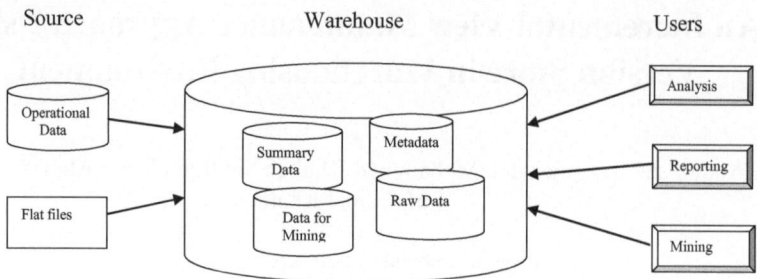

Fig. 1. Architecture of Data Warehouse

2 Existing Approaches for Maintaining Views

Various approaches have been introduced for maintaining the view in a warehouse environment.

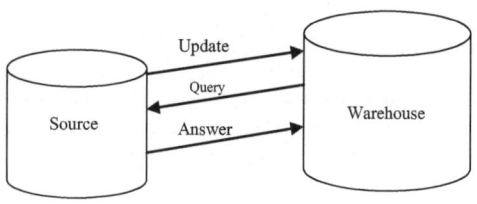

Fig. 2.

2.1 Basic Algorithm

At source:

   i. Updating at source (S ui) : execute ui;
      send ui to the warehouse;
      trigger event W ui at the warehouse.

  ii. Query at source (S qui): receive query qi;
      Let Ai=Qi[ssi]; (ssi is a current source state)
      Send Ai to warehouse;
      Trigger event W ansi to the warehouse

At warehouse:

   i. W upi : receive update ui;
      Let Qi = v(ui) ;
      Send Qi to the source;
      Trigger event S qui at the source

  ii. W ansi : receive Ai;
      Update view: $MV = MV + A;$

The above algorithm [3] states that:

1. When an update occurs at the source, it sends the update notification to the warehouse.
2. Warehouse receives the notification and sends back the query to the source about the update.
3. Source receives the query sent by the warehouse and returns the answer to that query.

2.2 RVAlgorithm

RV does not rely on incremental view maintenance approach. It is based on recomputation of materialized view from the scratch. In RV approach warehouse sends the Query to the source asking it to recompute the view from the scratch after certain number of updates. RV sends 2 messages for each update. The bytes transferred is much higher in RV than the relative algorithms. This degrades the performance of RV.

2.3 ECA (Eager Compensating Algorthm):

COLLECT = ϕ
W upi: receive Ui;
   Let Qi= v(Ui) – Σ Qj∈UQSQj(Ui)
   send Qi to the source;
   trigger event S qui at the source
W ansi: receive Ai;
   let COLLECT = COLLECT + Ai;
   if UQS = ϕ
   then { MV ←MV + COLLECT;
      COLLECT ← ϕ}
      else do nothing.

ECA is an incremental view maintenance algorithm. It is a method for fixing the view maintenance problem that occurs due to the decoupling between base data and the view maintenance manager at the warehouse. The key idea of the ECA algorithm is that it cannot rely on the state of the base information that are continuously being updated/modified by the sources. It must keep track of the updates received from the source and then filter out i.e., compensate any information that will duplicate the resulting queries. By subtracting (or adding) the results it knows that will (not) get in future queries, it will create an accurate end result for the view [3].

2.4 Lazy Approach

Lazy approach maintains the view in a lazy manner that relieves the updates of the maintenance overhead as in the incremental view maintenance approaches. View maintenance is postponed until the system has free cycles or it is referenced by any query. These free cycles are utilized for the view maintenance that relieves the updates and queries form the overhead. The updates are combined from different transactions into a single maintenance task. It also exploits row versioning. In lazy maintenance the updates do not maintain the view it just stores the required information so that the affected views can be maintained later. It actually uses system free cycles to maintain the views, in this no updates or queries pay for the maintenance task. But, in case the view is not up to date and query is sent over it, then the particular query has to pay for all part of the view maintenance and some delay also. However, it pays only the view maintenance that it uses and not for other views [2].

3 The Possible Solution

The maintenance of structural modifications in data warehouses is a crucial point for keeping track of structural modifications. We have proposed a solution in which we used the concept of some exiting approaches like version store [2]. Our solution is based upon incremental view maintenance utilizing the functionality of version store [2]. Assume a user updates some data in the source. In turn the source sends the update (Ui) notification to the ware house. Simultaneously, the copy of the updated data or that particular record(s) with the TXN number gets stored in the version store. The warehouse after receiving the notification prepares a query (Qi) and sends it back to the source. The source receives the query and return back the answer (Ai) to that query to the warehouse. This is the normal functioning of the system if the synchronization between the warehouse and source is maintained. But, in actual the scenario is completely different there is no synchronization between source and the warehouse as stated above in section 3. In our system we just tried to overcome the problems with the traditional approaches in section 2. We have described our detailed solution in the section 4.1 below.

3.1 Working of Our System

We have classified our work in the following steps and presented a flow diagram of the system in fig1.3.

Steps:

1. A user makes some change in the source (update, insertion, deletion etc).
2. The updated data (table) gets stored with TXN number in version store.
3. Source notifies the warehouse about the Update.
4. Warehouse checks the current sending Query (Qi).
    4.1. If it is null, it sends back the query to the warehouse goto 9.
    4.2. If not null, goto 5.
5. If the current Qi is equal to Ai goto 10.
6. If the current query Qi is not equal to Ai, then the updated data is fetched from version store.
7. The version store returns the answer by matching the TXN number, goto 4.
8. Warehouse receives the answer goto 5.

9. Source acknowledges the warehouse request and returns the answer to that query.
10. Warehouse updates the view with the received answer.
11. End.

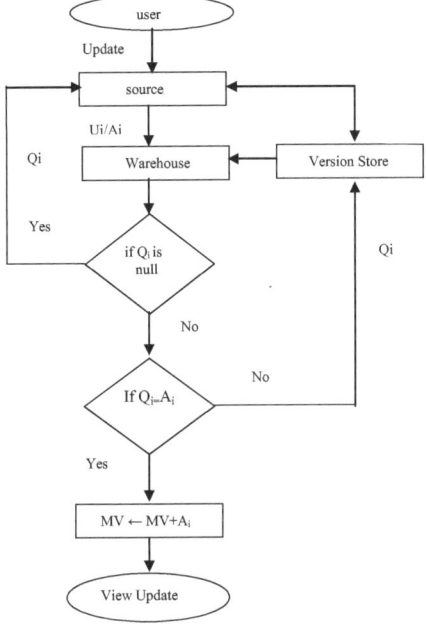

Fig. 3. Data Flow

3.2 View Manager

The main responsibility of View manager is to maintain the materialized views. The view manager keeps track of all the maintenance tasks for different views. It uses hash table to store an entry of a view that has an active maintenance task. With each entry a list is maintained with the transaction numbers (TXN) for the number of tasks that are in sorted order.

To update a view, view manager gets the maintenance task from the list and contacts the version store. The versions store has all the updated data with transaction numbers (TXN). The view manager matches the TXN numbers and retrieves the specific data from the version store and updates the view accordingly.

3.3 Version Store

Version store is storage of updated data [2]. It stores old versions of tables that have been updated. A transaction number TXN is given to each transaction in the table so as to maintain the versioning. These transaction numbers TXN are monotonically increasing with every transaction. Even each statement in a transaction is given a statement number STMTN. So that it is possible to figure out which statement belongs to which transaction. It maintains a version chain of the records so that it can recover the older versions of the records until they are available. Actually, this is not a persistent but the transient storage. It stores the older versions of tables for a short period of time. A version has to be kept only until all transactions that may require it have terminated. Then garbage collector reclaims the memory used by the stored data. Version tables greatly simplify the view maintenance tasks. If in the case of two or more then two consecutive updates at the source, the traditional systems are unable to get all the updated states from the source which develops the anomaly in the view maintenance.

In our system version tables helps view manager to overcome this anomaly. We explain this in our next section 4.3.

4 Performance Evaluation

In section 3 we have outlined several approaches from view maintenance in a warehousing environment i.e., re-computing view (RV), eager compensating algorithm (ECA). All these approaches provide consistent and non redundant materialized views at the warehouse. In the next section we are addressing the performance based on the messages transferred in maintaining a view. We intuitively compare RV and ECA with our proposed system.

4.1 Performance Based on Number of Messages

Let us suppose that there are n numbers of updates. In case of Recomputing View (RV) assume the warehouse sends a query to the source asking it to recompute the view after m number of updates where $m \leq n$. If we compute both query and answer messages the total numbers of messages are MV = (n / m) x 2 [3]. Therefore it shows that RV generates at least 2 messages if the view is recomputed once i.e., if m = n. Now suppose if the view is recomputed after every m number of updates where m < n then RV generates 2n number of messages. In ECA if there are n updates there are also n number of queries and the same number of answers, so there are 2n messages.

In our system if the updates are sufficiently spaced so that each query is processed before a new update occurs our solution works same as ECA i.e., 2n number of messages. In the worst case the total numbers of messages are higher than the RV and ECA as shown in the fig 1.3. The answer from the source is matched with the query sent if the match doesn't occur the answer is fetched from the version store. In this case the total numbers of messages transferred are 4m. So, the worst case is not desirable in our solution as the total transfers are much higher than the existing approaches.

5 Conclusion

Data warehousing is an emerging and already very popular technique used in many applications for retrieval and integration of data from autonomous sources. However, warehousing typically is implemented in an ad hoc way. We have shown that the standard algorithm for maintaining the materialized view at a warehouse can lead to anomalies and inconsistent modification to the views. the anomalies are due to the fact that view maintenance at the warehouse is decoupled from the updates at the data sources, and we cannot expect the data sources to perform sophisticated functions in support of view management. Consequently previously proposed view maintenance algorithm cannot be used in this environment.

We have presented our new approach that correctly maintains the materialized view in a ware housing environment. We have used the version store [2] to store the old states of the updated records with the transaction id, so as to figure out which update belongs to which transaction. This greatly helps in our new incremental view maintenance approach to overcome the anomalies in the previous standard approaches.

References

1. Gupta, A., Mumick, I.S. (eds.): Materialized Views: Techniques, Implementations and Appliacations. MIT Press, Cambridge (1999)
2. Zhou, J., Larson, P., Elmongui, H.G.: Lazy Maintenance of Materialized Views. In: Proceedings of the 33rd International conference on Very large data bases, Vienna, Austria (2007)
3. Zhuge, Y., Garcia-Molina, H., Hammer, J., Widom, J.: View Maintenance in a Warehousing Environment. In: Proceeding of SIGMOD Conference (1999)
4. Breibart, Y., Gracia-Molina, H., Silberschatz, A.: Overview of multi database transaction management. VLDB Journal 1(2), 181–239 (1992)
5. Agrawal, D., Abbadi, A., El, S.A., Yurek, T.: Efficient View Maintenance at Data Warehouses. In: The Proceedings of the ACM SIGMOD International Conference on Management of Data, Tucson, Arizona, USA (1997)
6. Gupta, A., Mumick, I.S.: Maintenance of Materialized Views: Problems, Techniques, and Applications. IEEE Bulletin of the Technical Committee on Data Engineering 18(2), 3–18 (1995)
7. Gupta, A., Mumick, I.S., Subrahmanian, V.: Maintaining views incrementally. In: Proceedings of the 1993 ACM SIGMOD International Conference on Management of Data, Washington, D.C, May 1993, pp. 157–166 (1993)
8. Yeung, G.C.H., Gruver, W.A.: Multiagent Immediate Incremental View Maintenance for Data Warehouses. IEEE Transaction on Systems, Man, and Cybernetics – Part A: Systems and Human 35(2) (March 2005)
9. He, H., et al.: Asymmetric Batch Incremental View Maintenance. In: The Proceedings of the 21st International Conference on Data Engineering ICDE (2005). IEEE, Los Alamitos (2005)
10. Labio, W.J., Yang, J., Cui, Y., Garcia-Molina, H., Widom, J.: Performance issues in incremental warehouse maintenance. In: Proc. 2000 Intl. Conf. Very Large Data Bases, Cairo, Egypt (September 2000)
11. Mohania, M., Samtani, S., Roddick, J., Kambayashi, Y.: Advances and Research Directions in Data-Warehousing Technology. Published in Australasian Journal of Information Systems 7(1) (1999)

12. Qin, B., Wang, S., Du, X.: Effective Maintenance of Materialized Views in Peer Data Management Systems. In: Proceeding of the First International Conference on Semantics, Knowledge, and Grid (SKG 2005). IEEE, Los Alamitos (2006); 0-7695-2534-2/05 ©
13. Blakeley, J.A., Larson, P.-A., Tompa, B.W.: Efficiently Updating Materialized Views. ACM, New York (1986); 0-89791-191-1/86/0500/0061

The Study of Synchronization Framework among Multi-datasets

Dongjin Yu and Wanqing Li

School of Computer, Hangzhou Dianzi University, 310018 Hangzhou, China
{yudj,liwanqing}@hdu.edu.cn

Abstract. Data from different autonomous systems should be kept consistent to fulfill the business collaboration. This paper presents the framework, called MDSIF, for data synchronization among multi-datasets based on messages. With the shared data center, MDSIF provides the coherent data snapshots for all related datasets by sending, receiving, and storing the shared data encapsulated in messages. Furthermore, it introduces MDSIF's synchronization mechanism in the formal way, which resolves data conflicts by setting attribute confidences and current value confidences respectively. According to this mechanism, those from the dataset with higher confidence value are allowed to overwrite those from the dataset with lower one. Finally, the case based on MDSIF is illustrated, showing that MDSIF needs only minimum manual intervention and is much more flexible, extendable and reliable, compared with the traditional trigger-based approach.

Keywords: Data Synchronization, Data Conflicts, Framework, Messages.

1 Introduction

The enterprise-level data center is usually constructed based on the data collected from multiple autonomous systems. Meanwhile, for the reason of business collaboration, different applications also need to be coupled with the underlying data swapping across system boundaries. In both cases, the values from different systems for the same entity may be inconsistent, simply because they are supervised under different domains and in different ways.

Data synchronization is an automated action to make the replicated data be consistent with each other and up-to-date. Through synchronization, the inconsistent, or dirty data, could be replaced with the right data timely and automatically. Since the data synchronization may involve large amount of data transferred under wide area network, it however tolerate moderate time delay.

The problem of data synchronization attracts many researchers' attention. The related topics include incremental replication methods, synchronization strategy, synchronization in mobile environments, synchronization performance and scalability, and so on. For instance, Li Xianmin et al. proposes an active differential

D. Ślęzak et al. (Eds.): DTA 2009, CCIS 64, pp. 17–25, 2009.

data synchronal model in which the change of the source data is listened actively and the update of the destination data is done synchronously through a monitor [1].

The conflict resolution of data synchronization is extremely important in mobile environments. In order to adapt synchronization to bandwidth and resource constraints, in [2], the concept of item relationships is adopted to synchronize only the items relevant to the user. Under some circumstances, the recent-data-win policy is adopted as the resolution rule in the mobile environment [3].

The manual intervention is usually unavoidable for resolving conflicts occurred in different datasets. However, Heumer G. achieves the automatic data exchange and synchronization by integrating an extra central knowledge component, which handles the incoming possibly conflicting changes and treats different data formats uniformly based on the declarative attribute representation concept [4].

Many focused on how to improve the performance and scalability for synchronizing large replicated datasets among distributed systems. Exploiting parallelism, omitting transfer of unnecessary metadata, synchronizing at a block level rather than a file level, using sophisticated compression methods are all feasible ways to achieve these goals [5].

Because data synchronization would bring about huge business profits, the industrial track has also developed many related products. Oracle, for instance, provides the synchronization mechanism based on Oracle Streams which have great advantages over traditional synchronous and asynchronous replication methods [6].

Different with the above mentioned ones, this paper gives a novel synchronization framework for multi-datasets based on messages. The framework configures shared data centers for data dumping, and resolves the conflicts with the help of Attribute Confidences and Current Value Confidences. It is most suitable under the relative poor networking environment with minimum manual intervention and could dump the swapping data temporally in case of network down.

The rest of the paper is organized in the following manner. Section 2 presents the framework architecture based on messages. Following the overall introduction of conflict resolving strategy introduced in Section 3, Section 4 examines the resolving methods in the formal way. The successfully implemented case is illustrated in Section 5. Finally, Section 6 provides concluding remarks and offers future research directions.

2   The Synchronization Framework

The star-like synchronization framework presented in this paper is named as the Message-based Data Synchronization and Integration Framework, or simply MDSIF (Fig. 1). Beside the star-like topology, the framework might otherwise adopt the peer to peer (P2P) architecture without any synchronization centers, which would avoid performance bottlenecks [7]. However, in P2P architecture, the number of data swapping paths would increase dramatically with more datasets involved, which eventually made the framework too complicated to administrate. Moreover, it would be hard to form the global data view which is usually required for enterprise-level decision support systems.

MDSIF is configured with several synchronizing centers, which are responsible for the data synchronizing and dumping tasks in their own administrating regions. The synchronization is fulfilled through the service buses which connect all the nodes of administrating centers represented by R, and participating datasets, represented by D in Fig. 1. Here, the nodes connected by the same bus belong to one group, called domain. Thus, MDSIF would be easily extended to make different domains coupled by cascading service buses.

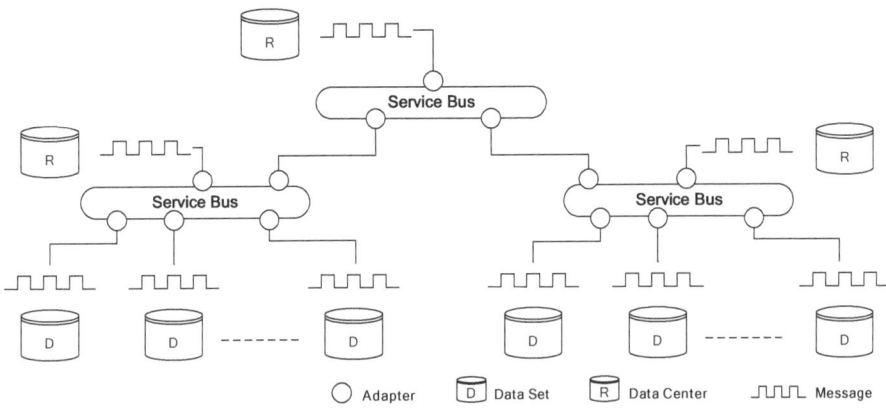

Fig. 1. The Cascading Synchronization Framework Based on Messages

Due to the networking instability in the wide area distributed environment, MDSIF introduces the asynchronous communication model based on messages. The message itself encapsulates both the control information and the data. When asked to synchronize the data, the message sender sends the data and continues to execute its own task with no need to wait for the response from message receiver. On the other hand, the message receiver is not required to handle the request immediately once it receives the data. The messages would be kept in the message queue until they are consumed or handled eventually. Thus, the data integrity and consistence would be fully guaranteed in case of network breakdown.

MDSIF provides the following two data synchronizing modes.

1) Publish-Subscribe Mode

The data provider does not send the data directly to the receiver. Instead, the data are published under certain topics. The MOM (Message Oriented Middleware) will propagate the data to the receiver which has subscribed it via the service bus. In this mode, the sending of synchronized data is initiated by the data provider.

2) Request-Response Mode

The node demanding data requests the MOM for the data synchronization service. The latter then interacts with the corresponding data provider via service buses. Once the data are obtained, the MOM sends it back to the demanding node. In this mode, the sending of synchronization data is initiated by the demand side.

3 Conflict Resolving

In distributed complicated environments, data conflicts usually occur when the values from two or more participating datasets differ with each other although they represent the same entity. Data conflicts bring the ambiguity, which dramatically decreases the data quality.

Traditionally, database triggers are adopted to resolve data conflicts. Once the data source catches the operations of value insertion, deletion or modification, it executes the predefined triggers to redo the same operations in the targets through database links. However, the trigger-based approach only allows the execution of fixed and unidirectional synchronization, which transfers the data in the predefined source to the predefined target. In fact, however, data in the datasets other than the predefined source may need to be synchronized with the data in target, when the data in the predefined source are temporary absent or lost. Under other circumstances, the data in source are even allowed to be replaced by the data in target if the source data are null or could be simply ignored, which leads to the bidirectional synchronization.

In MDSIF, the data conflict might be resolved by predefined data confidence values which denote the extent of data reliability. Different with the traditional trigger-based approach, MDSIF does not predefine data sources or data targets. For the same attribute of same entity, those from the dataset with higher confidence value are all allowed to overwrite those from the dataset with lower one (Fig. 2).

There are two kinds of confidences in MDSIF:

1) AC: Attribute Confidences

Every shareable attribute in each dataset has its own Attribute Confidence to indicate the reliability of all values created from this dataset for this attribute. Attribute Confidences are determined before the synchronization happens by the data managers who are familiar with and in charge of the data's business value. Attribute Confidences are stored in the Attribute Confidence Table. Moreover, Attribute Confidences are fixed and could not be changed after setting.

2) CVC: Current Value Confidences

The Current Value Confidence represents the present confidence of each shareable value. Since the Current Value Confidence is set automatically according to the Attribute Confidence of the corresponding attribute in its original dataset, it may be different with its own Attribute Confidence. Current Value Confidences are stored in Current Value Confidence Table. Different with Attribute Confidences, Current Value Confidences may change during the whole synchronization process.

4 Synchronization Methods

This section describes the method in detail to get data consistent through synchronizing unmatched values among different datasets in MDSIF. For simplicity, the method described here is limited in one domain.

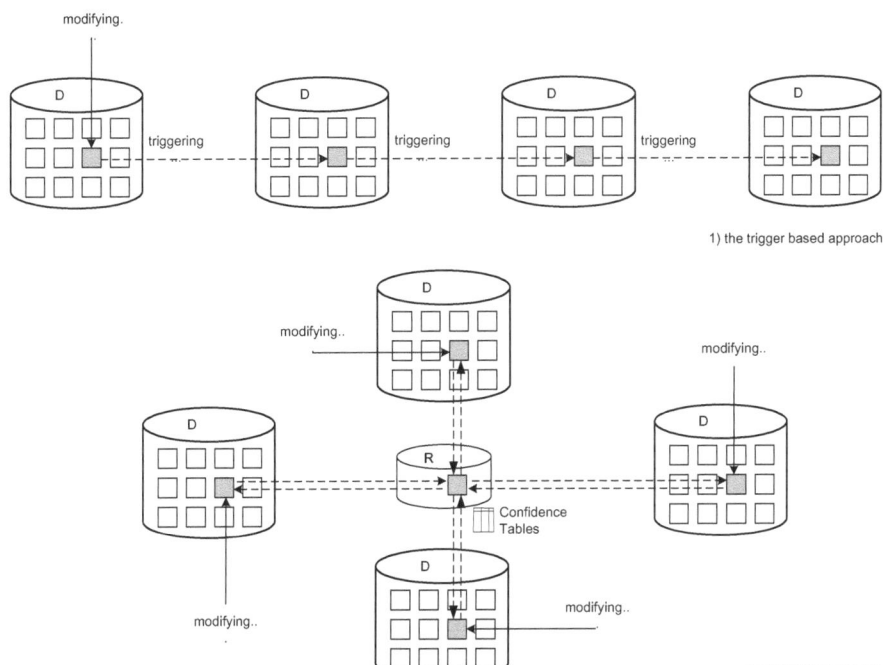

1) the trigger based approach

2) the MDSIF approach

Fig. 2. The Synchronization Flows in the Trigger-based Approach and the MDSIF Approach

Definition 1. A distributed *data synchronization framework* is denoted as a quadruple E, i.e. $E = \{D, U, R, T\}$, in which:

$D = \{D_1, D_2, ..., D_n\} n \geq 2$, is the combination of all participating datasets, named as the *synchronization datasets*, or *simply datasets*.

$U = \{u_1, u_2, ..., u_k\}$, $k \geq 1$, is the *attributes' set* whose values should be kept consistent in D.

$R = R^{u_1} \cup R^{u_2} ... \cup R^{u_k}$, is the *shared dataset center* formed by synchronizing all the values of *attribute set* in U.

$T = \{T_0, T_1, ..., T_{m-1}\}$, is the set of *synchronization rules*.

Definition 2. For the attribute u, the extent of reliability of dataset D_i compared with other datasets in E is called the *Attribute Confidence* of D_i for attribute u, represented as C_i^u, where $0 \leq C_i^u \leq 1$. The higher the reliability, the bigger C_i^u.

Definition 3. For the shared value r, the extent of reliability of dataset D_i compared with other datasets in E is called the *Current Value Confidence* of r, represented as V_i^r where $0 \le V_i^r \le 1$. The higher the reliability, the bigger V_i^r.

Definition 4. If $C_i^u = 1$, then D_i is called the *dataset with full confidence*, or *determined dataset*, in E for attribute u, denoted by M^u.

Definition 5. If $C_i^u = 0$, the dataset D_i is called the *dataset with no confidence*, or *ignored dataset*, in E for attribute u, denoted by I^u.

With the above definitions, the synchronization method in the framework E could be described as following.

Step 1) Sort *Attribute Confidences*

For each attribute u, reorder all the datasets in D, except the determined dataset M^u, according to their Attribute Confidences from high to low. Assume the result is:

$$D^T =< D_{i_1}, D_{i_2}, \ldots, D_{i_{n-1}} >, C_{i_1}^u \ge C_{i_2}^u \ge \cdots \ge C_{i_{n-1}}^u \tag{T0}$$

Note there is always only one *determined dataset* at most for each shared attribute.

Step 2) Construct the *shared dataset center* R

Let R^u denote the shared dataset for attribute u, then:

$$R^u = \begin{cases} (r^u, v^r) | r^u \in M^u \cup \left(D_{i_1}^u - M^u \right) \cup \left(D_{i_2}^u - D_{i_1}^u - M^u \right) \\ \cup \ldots \cup \left(D_{i_{n-1}}^u - D_{i_{n-2}}^u - \cdots - D_{i_1}^u - M^u \right), \\ v^r = \max(C_i^u | r^u \in D_i^u) \end{cases} \tag{T1}$$

where v^r represents the *Current Value Confidence* for value r, i.e. the highest *Attribute Confidence* among all *synchronization datasets* containing r^u.

In particular, (T1) could be simplified as (T2), which simply means that all datasets except the *determined dataset* are *ignored datasets* for attribute u.

$$R^u = \{(r^u, v^r) | r^u \in M^u, v^r = 1.0\}, C_{i_1}^u = C_{i_2}^u = \cdots = C_{i_{(n-1)}}^u = 0 \tag{T2}$$

Thus the whole shared dataset center R could be constructed as following:

$$R = \cup R^u, u = u_1, u_2, \ldots, u_k \tag{T3}$$

Step 3) Duplicate the shared data

Copy the shared data from the shared dataset center R to all related datasets via messages sending and receiving:

$$D_i^{u_j} = \{r | r \in R^{u_j}\}, i = 1, 2, \ldots, n; j = 1, 2, \ldots, k \qquad (T4)$$

Step 4) Synchronize on demand

Subsequent synchronization occurs when any data modification may introduce the inconsistency after the initial round of synchronization illustrated in Step 2) and 3). For the modification of existing value r of attribute u in D_i :

Step 4.1) If $C_i^u < V_i^r$, then reject the modification.

Step 4.2) If $C_i^u \geq V_i^r$

Step 4.2.1) modify r in D_i.

Step 4.2.2) modify r in the *shared dataset center* R, and change its *Current Value Confidences* to C_i^u.

Step 4.2.3) modify r in the *synchronization datasets* $D_1, D_2, \ldots, D_j, \ldots, D_n, j \neq i$.

5   Case Studies

Fig. 3 illustrates the Attribute Confidence Table and Current Value Confidence Table for the case conducted in the labor administration field. There are 3 separate systems altogether, i.e. the Social Security System (Node #1), the Professional Training System (Node #2) and the Employment System (Node #3) respectively. Here, the

1) Attribute Confidence Table

Attribute	Node ID.	AC
PersionID	#1	1.0
Birthdate	#1	*1.0*
Birthdate	#2	*0.8*
Birthdate	#3	*0.5*
Degree	#1	0.8
Degree	#2	1.0
Income	#1	0.5
Income	#3	1.0

AC: Attribute Confidence

2) Current Value Confidence Table

Record ID.	Attribute	CVC
1001	PersonID	1.0
1001	*Birthdate*	*0.8*
1001	Degree	0.8
1001	Income	0.5
1002	PersonID	1.0
1002	Birthdate	1.0
1002	Degree	1.0
1002	Income	1.0

CVC: Current Value Confidence

Fig. 3. The Attribute Confidence Table and Current Value Confidence Table for the Case

updatable status could be deduced according to the related values in the Attribute Confidence Table and Current Value Confidence table. For example, the value of birthdate (denoting the individual's birth date) for Record 1001 could be replaced with the new value created from Node #1 or Node #2, but not the value from Node #3, because its Current Value Confidence (0.8) is bigger than the corresponding Attribute Confidence in Node #3 (0.5), but smaller than or equal to that in Node #1 (1.0) and Node #2 (0.8).

The IBM Websphere MQ [8] is adopted in the case as the Message Oriented Middleware (MOM) to send, receive and store the swapped data encapsulated in messages. Currently, the datasets are kept in Oracle databases in 3 different nodes, while both the Attribute Confidence table and the Current Value table are gathered in the center node.

6 Conclusions

The message-based data synchronization framework, MDSIF, could ensure the reliable synchronization by message buffering under unguaranteed networking condition. In addition, it could automatically resolve the data conflicts via the labeled confidence values. Different with the traditional trigger-based approach, MDSIF is more flexible, extendable, reliable and easier to provide the global data view.

Since MDSIF configures the Attribute Confidence tables and Current Value Confidence tables in the remote shared data centers, the overhead of confidence setting and querying is somewhat expensive. Future research could include the consideration of distributing the above confidence tables to different datasets for local access in order to reduce the synchronization time.

Acknowledgments. The work is supported by the Foundation of Zhejiang Provincial Key Science and Technology Projects (No. 2008C11099-1), the Science Foundation of the Hangzhou Dianzi University (No. KYS055608069, No. KYS055608057), and the Natural Science Foundation of Zhejiang province of China (No. Y6090312).

References

1. Xianmin, L., Hui, W.: The Model of the Active Differential Data Synchronization for the Heterogeneous Data Source Integration Systems. In: The First International Symposium on Information Technologies and Applications in Education, pp. 572–574 (2007)
2. Koskimies, O.: Using data item relationships to adaptively select data for synchronization. In: Kutvonen, L., Alonistioti, N. (eds.) DAIS 2005. LNCS, vol. 3543, pp. 220–225. Springer, Heidelberg (2005)
3. Lee, Y.S., Kim, Y.S., Choi, H.: Conflict resolution of data synchronization in mobile environment. In: Laganá, A., Gavrilova, M.L., Kumar, V., Mun, Y., Tan, C.J.K., Gervasi, O. (eds.) ICCSA 2004. LNCS, vol. 3044, pp. 196–205. Springer, Heidelberg (2004)
4. Guido, H., Malte, S., Erich, L.M.: Automatic data exchange and synchronization for knowledge-based intelligent virtual environments. IEEE Virtual Reality, 43–50 (2005)

5. Schutt, T., Schintke, F., Reinefeld, A.: Efficient synchronization of replicated data in distributed systems. In: Sloot, P.M.A., Abramson, D., Bogdanov, A.V., Gorbachev, Y.E., Dongarra, J., Zomaya, A.Y. (eds.) ICCS 2003. LNCS, vol. 2657, pp. 274–283. Springer, Heidelberg (2003)
6. Hao, Y., Xing-chun, D., Guo-quan, J.: Research on Data Synchronization in Oracle Distributed System. In: 2008 International Seminar on Future Information Technology and Management Engineering, pp. 540–542 (2008)
7. Endo, S., Miyamotot, T., Kumagait, S., Fujii, T.: A Data Synchronization Method for Peer-to-Peer Collaboration Systems. In: International Symposium on Communications and Information Technologies, pp. 368–373 (2004)
8. International Business Machines Corp,
   http://www-01.ibm.com/software/integration/wmq/

Clustering News Articles in NewsPage.com Using NTSO

Taeho Jo

School of Computer and Information Engineering, Inha University
230 Yonghyundong Namgu Incheon 402-751 South Korea
tjo018@inha.ac.kr
http://tjo018.inha.ac.kr

Abstract. In this research, the NTSO (Neural Text Self Organizer) is proposed as the approach to text clustering. It is required to encode documents into numerical vectors for using a traditional approach to text clustering. The two main problems, huge dimensionality and sparse distribution are caused by encoding so. The idea of this research is to encode documents into string vectors and use the NTSO as the approach to text clustering. As the empirical validation, we will compare the NTSO with other text clustering approaches with respect to the speed and the performance.

Keywords: Neural Text Self Organizer, Text Clustering.

1   Introduction

Text clustering refers to the process of segmenting a group of various documents into subgroups of content based similar documents. The proximity and clustering are the main criteria for executing the text clustering [1]. The proximity indicates the criteria for measuring the similarity between two documents. The second criteria, 'clustering' means the decision whether a document is included in a given cluster. Text clustering is necessary as a fundamental technique for organizing documents based on their contents.

In order to use a traditional algorithm for text clustering, documents should be encoded into numerical vectors. The two main problems, huge dimensionality and sparse distribution, are caused by encoding documents so. The first problem, huge dimensionality, means the requirement of many features for doing the text clustering robustly, since each feature has very weak coverage. The second problem, sparse distribution means that zero values occupy dominantly in each numerical vector. This problem degrades the performance of text clustering because the discrimination among numerical vectors is lost.

The first point of the idea of this research is to encode documents into string vectors, instead of numerical vectors. A string vector refers to a finite ordered set of words or strings. In each string vector, a string or a word becomes an element. Grammatical, statistical, and posting properties of words may be defined as

D. Ślęzak et al. (Eds.): DTA 2009, CCIS 64, pp. 26–33, 2009.

features of string vectors. This point becomes the complete solution to the two main problems which are inherent in encoding documents into numerical vectors.

The second point of the idea is to apply the NTSO to text clustering as a string vector based approach. The proposed neural network consists of two layers: the input layer and the competitive layer; the architecture is same to the Kohonen Networks. However, note that the proposed neural network is essentially different from the Kohonen Networks in that input vectors and weight vectors are given as string vectors, instead of numerical vectors. The two operations on string vectors involved in the learning process of the proposed neural network will be defined in this research. Hence, the proposed neural network deals with string vectors unlike traditional neural networks.

As the effect of this research, the two main problems are completely avoided by encoding documents into the string vectors and applying the new type of neural network, called NTSO (Neural Text Self Organizer). The first problem, the huge dimensionality, is completely solved by encoding documents more compactly since each feature of string vectors is applicable to every document. By reducing the size of input data, it is expected to improve the speed of clustering. The second problem, the sparse distribution, is not inherent in encoding documents into string vectors since each string vector has a single value dominantly with very little probability. It is expected for the avoidance of the two problems to improve the performance of text clustering, since more discrimination among structured data representing documents are obtained.

2   Previous Works

The single pass algorithm may be considered, as a typical approach to text clustering. It was initially created by Sylester and Seth in 1995 [2]. It was initially applied to text clustering by Papka and Allan in 1998 [3]. It has ever subsequently used for text clustering by Hatzilbassilogou et al in 2000 [4]. Note that the single pass algorithm has its very good speed but its very poor performance.

The K means algorithm may be considered as another approach to text clustering. It was initially created by Hartigan and Wong in 1979 [5]. It was initially applied to text clustering by Beil et al in 1994 [6]. It was subsequently used for text categorization by Larsen and Awone in 1999 [7]. Compared with the single pass algorithm, the k means algorithm have its better clustering performance but lower speed.

The Kohonen Networks may be considered as the typical approach to text clustering. It was initially created by Kohonen in 1982 [8]. It was initially used for implementing the text clustering system called WEBSOM by Kaski et al in 1998 [9]. In 2000, Kohonen et al modified the system to improve the speed [10]. Like the k means algorithm, it has its better performance but it is slower than the single pass algorithm.

The EM algorithm exist as a framework of clustering algorithms rather than a specific algorithm. The framework was initially proposed by Dempster in 1977 [11]. Based on the EM algorithm, in 1998, a specific clustering algorithm was created and applied to text clustering by Ambroise and Govaert in 1998 [12]. Subsequently, in 2000, more specific clustering algorithms based on EM algorithm were proposed and applied to text clustering by Vinokourov and Girolami [13]. Therefore, many specific clustering algorithms are available based on the EM algorithm.

As the previous solution to the two main problems in encoding documents into numerical vectors for text clustering, string kernel may be considered. It was initially proposed by Lodhi et al in 2002 [14]. It was used as a kernel function of SVM (Support Vector Machine) and computes syntactic similarity between two long strings. Since it uses two raw documents without encoding them into numerical vectors, the two main problems were avoided completely. However, the solution was not for text clustering, but for text categorization, and furthermore, it failed to improve the performance of text categorization.

3   String Vectors

3.1   Definition of String Vector

A string vector refers to an ordered finite set of strings. It is denoted by $[s_1, s_2, ..., s_d]$, where s_i is a string given as an element and d refers to the fixed number of elements. The number of elements of the string vector, d, is called the dimension of the string vector. Since the elements of the string vector are ordered, two string vectors with their identical elements but their different orders are treated as different ones. [computer information data], [business company industry], and [hardware machine part] are examples of three dimensional string vector.

3.2   Similarity Matrix

The similarity matrix is denoted as follows:

$$\begin{pmatrix} s_{11} & s_{12} & \cdots & s_{1N} \\ s_{21} & s_{22} & \cdots & s_{2N} \\ \vdots & \vdots & \ddots & \vdots \\ s_{N1} & s_{N2} & \cdots & s_{NN} \end{pmatrix}$$

In the above matrix, N indicates the total number of words, and the similarity matrix is a $N \times N$ matrix. In the similarity matrix, its columns and rows correspond to words; the $i$the column and the ith row correspond to the identical word. The entry of the similarity matrix, s_{ij} indicates the semantic similarity between the word which corresponds to $i$the column or the ith row and the word which corresponds to $j$the column or the jth row, and it is computed by equation (1),

$$s_{ij} = \frac{2 \times df(w_i, w_j)}{df(w_i) + df(w_j)} \tag{1}$$

where in the corpus, $df(w_i, w_j)$ indicates the number of documents including both words, w_i and w_j, $df(w_i)$ indicates the number of documents including the word, w_i, and $df(w_j)$ indicates the number of documents including the word, w_j. Therefore, in the given corpus, the more documents including both words, the higher the semantic similarity between the words is.

3.3   Average Semantic Similarity between Two String Vectors

Two string vectors are denoted as follows:

$$\mathbf{str_1} = [str_{11}, str_{12}, ..., str_{1d}], \mathbf{str_2} = [str_{21}, str_{22}, ..., str_{2d}]$$

The similarities of 1:1 pairs of two string vectors denoted by $sim(str_{11}, str_{21})$, $sim(str_{12}, str_{22})$, ..., $sim(str_{1d}, str_{2d})$ are computed by getting elements crossing in the rows and columns corresponding to the given strings or words. If there is no such word which corresponds to its row or column, the similarity of the word becomes zero. The average semantic similarity between the two string vectors is computed, using equation (2).

$$sim(\mathbf{str_1}, \mathbf{str_2}) = \frac{1}{d} \sum_{i=1}^{d} sim(str_{1i}, str_{2i}) \qquad (2)$$

Since semantic similarities of one to one pairs are given as normalized values, the average value is computed as a normalized value.

The properties of the average semantic similarity between two string vectors are as follows:

- The average semantic simialrity between two identical string vectors is 1.0, since all one to one pairs of elements are 1.0.
- The associative law is applicable to this operation as expressed

$$sim(\mathbf{str_1}, \mathbf{str_2}) = sim(\mathbf{str_2}, \mathbf{str_1})$$

  because the semantic similarity is given as a symmetry matrix.
- This operation involves only two string vectors, since the similarity matrix is given as a two dimensional array.
- The operation generates a normalized value, since only normalized values are averaged.

3.4   Inter-word Set

The inter-words between two input words refer to those which are semantically relevant to the given input words. For example, if 'computer' and 'hardware' are given as the input words, CPU, memory, board, and hard disk are inter-words of the two words. The inter-words are formally defined as those which have their higher semantic similarities than the similarity between the two input words. If the words, str_i and str_j are the input words, we obtain $sim(str_i, str_j)$ as the similarity between the two words from the similarity matrix. From the similarity matrix, we can build the two sets of words which are satisfying $I_1 = \{str|sim(str, str_i) \geq sim(str_i, str_j)\}$ and $I_2 = \{str|sim(str, str_j) \geq sim(str_i, str_j)\}$, and get the inter-word set by intersecting the two sets: $I = I_1 \cap I_2 = \{str|sim(str, str_i) \geq sim(str_i, str_j) \land sim(str, str_j) \geq sim(str_i, str_j\}$.

Now, let's assume that the two string vectors are given:

$\mathbf{str}_i = [str_{i1}, str_{i2}, \ldots, str_{id}]$ and $\mathbf{str}_j = [str_{j1}, str_{j2}, \ldots, str_{jd}]$. We obtain the semantic similarity $sim(str_{ik}, str_{jk})$, element by element from the similarity matrix. We make inter-words sets as many as the elements, $\mathbf{I}_k = \{str|sim(str, str_{ik}) \geq sim(str_{ik}, str_{jk}) \land sim(str, str_{jk}) \geq sim(str_{ik}, str_{jk}\}$ A random element of the set, $\mathbf{A}$ is denoted by $rand(\mathbf{A})$. Therefore, the string vector, $\mathbf{I} = [rand(\mathbf{I}_1), rand(\mathbf{I}_2), \ldots, rand(\mathbf{I}_d)]$.

Let's consider the complexity of this operation. It is assumed that the dimension of the string vector is d, and the similarity matrix is a $N \times N$ square matrix. It costs the complexity $O(N)$ to trace elements in the similarity matrix for making the two sets, $\mathbf{I}_k = \{str|sim(str, str_{ik}) \geq sim(str_{ik}, str_{jk})\}$ and $\mathbf{I}_k = \{str|sim(str, str_{jk}) \geq sim(str_{ik}, str_{jk})\}$, and intersects them with each other. In order to make a string vector, $\mathbf{I} = [rand(\mathbf{I}_1), rand(\mathbf{I}_2), \ldots, rand(\mathbf{I}_d)]$, it costs the complexity, $O(Nd)$. Although it is linear complexity to the size of the similarity matrix, it takes very much time for executing the operation, because the value of N is too big.

4   NTSO

The architecture of the NTSO is illustrated in figure 1. The input layer receives an input vector given as a string vector and the number of nodes in the layer is consistent with the dimension of string vectors. Each node of the competitive layer corresponds to each cluster. The weight vectors between the two layers become prototypes of clusters given as string vectors. Although the NTSO follows the Kohonen Networks in terms of its architecture and learning process, the two neural networks are essentially different from each other in that the weight vectors and input vectors exist as string vectors.

The first step of the learning process is the initialization of weight vectors; there are three ways for initializing the weight vectors. The first way is to select some representative string vectors at random among training examples. The second ways is to collect all elements of string vectors from the training examples and select some of the collected elements at random. The third way is to retrieve

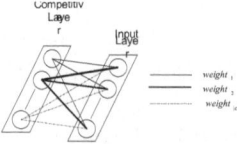

Fig. 1. The Architecture of the NTSO

words randomly from a particular corpus which is given, separated from the training examples. The initialized weight vectors signify the initial prototypes of clusters.

After initializing the weights, the optimization of weight vectors is started. Whenever an input vector is given, the node of the competitive layer whose weight vector is closest to the input vector is selected as the winning node. The similarity between the input vector and a weight vector is computed by the first operation, called the average semantic similarity. The weight vector connected to the winning node is replaced by the inter-string vector between the original weight vector and the input vector. The selection of the winning node and the update of the weight vector are applied to all input vectors, and they are iterated until the convergence of the weight vectors.

5   Experiments and Results

The partition of the test bed, NewsPage.com into the training and test set is illustrated in table 1. The test bed is given as a small collection of news articles for entering the first evaluation, and its source is the web site, www.newspage.com [1]. The collection was built by copying and pasting news articles individually as the plain text files. In the test bed, the five categories and the 1,200 news articles are available. The collection of news articles is partitioned into the training and test set by the ratio 7:3, as shown in table 1.

Table 1. Training and Test Set in the Test Bed: NewsPage.com

Category Name	Training Set	Test Set	Total
Business	280	120	400
Health	140	60	200
Law	70	30	100
Internet	210	90	300
Internet	140	60	200
Total	840	360	1200

The test bed is partitioned into various sized groups. Among the 20 categories, the four clearly different categories were selected. The ten variable sized subgroups are built: the smallest one has 100 documents, and the largest one has 1,000 documents. For each subgroup, documents are allocated, maintaining the balance over the four categories. We will observe the clustering performance and speed of the involved four approaches: the single pass algorithm, the k means algorithm, the Kohonen Networks, and the proposed approach by increasing the size by 100 from 100 sized group to 1,000 sized one.

Figure 2 illustrates the results from comparing the four approaches with respect to the clustering performance. In figure 3, the x-axis of the line graph

---

[1] The test bed was built in 2002, but the web site was recently closed.

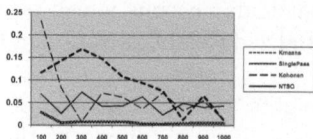

Fig. 2. The Comparison of Clustering Algorithms with Respect to the Clustering Performance

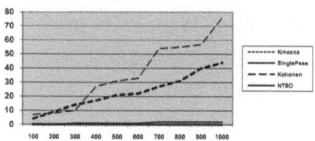

Fig. 3. The Comparison of Clustering Algorithms with Respect to the Clustering Speed

indicates the size of each group which means the number of documents contained in the group, and the y-axis of the line graph indicates the value of the clustering index which was adopted as the evaluation measure. The evaluation measure is described in the literature [15] and it is based on the intra-cluster similarity and the inter-cluster similarity. Among the four approaches, the single pass algorithm has the poorest clustering performance compared with other approaches, and the proposed one is comparable to the Kohonen Networks and the k means algorithm.

Figure 3 illustrates the results from comparing the involved approaches with respect to the clustering speed. In figure 4, the x-axis is same to that in figure 2, and the y-axis indicates the seconds which are taken for clustering each group. The single pass algorithm is fastest among the four approaches, and the k means algorithm is slowest. The proposed algorithm is comparable to the single pass algorithm with respect to the clustering speed. Therefore, the proposed algorithm is judged as the practical approach to text clustering with respect to both the clustering performance and speed as shown in figure 2 and 3.

6   Conclusion

Let's consider the contributions of this research. For first, the practical approach to text clustering was proposed in terms of the clustering performance and speed as shown in figure 2 and 3. For second, the two main problems, the huge dimensionality and the sparse distribution, were completely solved by encoding documents into alternative representations to numerical vectors. For third, the unsupervised neural network, NTSO (Neural Text Self Organizer) where the input vectors and weight vectors were as string vectors, was created. The string vectors representing documents are more transparent than the numerical vectors for guessing contents from the representations.

References

1. Halkidi, M., Batistakis, Y., Vazirgiannis, M.: On Clustering Validation Techniques. Journal of Intelligent Information Systems 17(2), 107–145 (2001)
2. Sylwester, D., Seth, S.: A trainable, singlepass algorithm for column segmentation, Technical Report UNL-CSE-95-003 of the Departement of Computer Science and Engineering at University of Nebraska-Lincoln (1995)
3. Papka, R., Allan, J.: On-Line New Event Detection using Single Pass Clustering, Technical Report UM-CS-1998-021 of the Department of Computer Science at University of Massachusetts (1998)
4. Hatzivassiloglou, V., Gravano, L., Maganti, A.: An Investigation of Linguistic Features and Clustering Algorithms for Topical Document Clustering. In: The Proceedings of 23rd SIGIR, pp. 224–231 (2000)
5. Hartigan, J.A., Wong, M.A.: A K-Means Clustering Algorithm. Applied Statistics 28(1), 101–108 (1979)
6. Beil, F.F., Ester, M., Xu, X.: Frequent term-based text clustering. In: The Proceedings of the eighth ACM SIGKDD international conference on Knowl- edge discovery and data mining, pp. 436–442 (1994)
7. Larsen, B., Aone, C.: Fast and Effective Text Mining Using Linear-time Doc- ument Clustering. In: The Proceedings of the fifth ACM SIGKDD international conference on Knowledge discovery and data mining, pp. 16–22 (1999)
8. Kohonen, T.: Self Organized Formation of Topologically Correct Feature Maps. Biological Cybernetics 43, 59–69 (1982)
9. Kaski, S., Honkela, T., Lagus, K., Kohonen, T.: WEBSOM-Self Organizing Maps of Document Collections. Neurocomputing 21, 101–117 (1998)
10. Kohonen, T., Kaski, S., Lagus, K., Salojarvi, J., Paatero, V., Saarela, A.: Self Organization of a Massive Document Collection. IEEE Transaction on Neural Networks 11(3), 574–585 (2000)
11. Dempster, A.P., Laird, N.M., Rubin, D.B.: Maximum Likelihood from In- complete Data via EM algorithm. Journal of the Royal Statistics Society, Series B 39(1), 1–38 (1977)
12. Ambroise, C., Govaert, G.: Convergence of an EM-type algorithm for spatial clustering. Pattern Recognition Letters 19(10), 919–927 (1998)
13. Vinokourov, A., Girolami, M.: A Probabilistic Hierarchical Clustering Method for Organizing Collections of Text Documents. In: The Proceedings of 15th International Conference on Pattern Recognition, pp. 182–185 (2000)
14. Lodhi, H., Saunders, C., Shawe-Taylor, J., Cristianini, N., Watkins, C.: Text Classification with String Kernels. Journal of Machine Learning Research 2(2), 419–444 (2002)
15. Jo, T., Lee, M.: The Evaluation Measure of Text Clustering for the Variable Number of Clusters. In: Liu, D., Fei, S., Hou, Z., Zhang, H., Sun, C. (eds.) ISNN 2007. LNCS, vol. 4492, pp. 871–879. Springer, Heidelberg (2007)

Categorizing News Articles Using NTC without Decomposition

Taeho Jo

School of Computer and Information Engineering, Inha University
tjo018@inha.ac.kr

Abstract. In this research, we attempt to apply the NTC (Neural Text Categorizer) to the text categorization without decomposing it into binary classifications. Because a single classifier has its very weak robustness to the entire text categorization, it is usually decomposed into binary classifications as many as categories. However, it requires to rearrange and relabel the given training examples with positive or negative labels for decomposing the text categorization. The task of this research is to apply the NTC to the text categorization without the decomposition and validate its feasibility. Therefore, we will compare the NTC with other approaches in the text categorization in the environment where the text categorization is not decomposed and validate that the NTC is practical tool for implement a light version of text categorization system.

1 Introduction

Text categorization refers to the automatic assignment of one or some of the predefined categories to each text. As the preliminary tasks, manually, we must predefine a list of categories and prepare sample labeled documents. Using the sample labeled documents, classification capacities are built, as the learning process. Using the classification capacities, unseen documents are classified automatically. Therefore, the text categorization system runs with the three steps. It requires the decomposition of the entire text categorization into binary classifications as many as categories for applying a machine learning algorithm to the task. Especially, in using the SVM for the task, the decomposition is mandatory. Note that it costs very much for the decomposition. The sample labeled documents should be arranged and each of them should be labeled again with the positive or negative class. In order to save the cost for the decomposition, we attempt to adopt and apply the NTC (Neural Text Categorizer) in this research.

If the feasibility is guaranteed, there are advantages in applying a machine learning algorithm to the text categorization without the decomposition. The first advantage is that we need not to relabel the given sample labeled documents. The second advantage is that only a classifier is enough for carrying out the entire text categorization. The third advantage is that the potential possibility of implementing a text categorization system as a light version is provided. However, note that the performance is not good as the case with the decomposition.

D. Ślęzak et al. (Eds.): DTA 2009, CCIS 64, pp. 34–40, 2009.

Let's consider the potential possibility of using the NTC for the entire text categorization without the decomposition. Over different categories, the distributions of words are significantly different. Each learning node in the proposed neural network has its own unique table which corresponds to each category. The CSVs (Categorical Score Values) are decided by the different tables within the learning nodes. Therefore, it is possible to construct distinguishable tables in multiple categories as well as two classes in the NTC.

This research provides the potential possibility of implementing a text categorization system as its light version which is suitable for the ubiquitous or embedded environments. Empirically, the NTC was already validated as the practical approach by comparing it with other approaches in the experiments with the decomposition. However, with the decomposition, the classifiers as many as categories were required. The decomposition leads the system to be very heavy. Without the decomposition, the system becomes light, because only one classifier is required in the system.

This article consists of the six sections including this section. In section 2, we will explore the previous research relevant to this research. In section 3 and 4, we will describe the definition and operations of string vectors and the NTC, respectively. In section 5, we will validate empirically the feasibility of the NTC in the text categorization without the decomposition. In section 6, as the conclusion of this article, we will mention the significance and remaining tasks of this research.

2 Previous Works

The KNN may be considered as a typical and popular approach to text categorization [1]. The KNN was initially created by Cover and Hart in 1967 as a genetic classification algorithm [2]. It was initially applied to text categorization by Massand et al in 1992 [3]. KNN was recommended by Yang in 1999 [4] and by Sebastiani in 2002 [1] as a practical approach to text categorization. Therefore, the KNN has been aimed as the base approach in other literatures as the base approach [1].

The Naive Bayes may be considered as another approach to text categorization. It was initially created by Kononenko in 1989, based on Bayes Rule [5]. Its application to text categorization was mentioned in the textbook by Mitchell in 1997 [6]. Assuming that the Naive Bayes is the popular approach, in 1999, Mladenic and Grobelink proposed and evaluated feature selection methods [7]. The Naive Bayes has been compared with other subsequent approaches in text categorization [9] [10].

Recently, the SVM was recommended as the practical approach to text categorization [9] [10]. It was initially introduced in her magazine article by Hearst in 1998 [8]. In the same year, it was applied to text categorization by Joachims [9]. It was adopted as the approach to spam mail filtering as a practical instance of text categorization in 1999 by Druker et al [10]. Furthermore, the SVM is popularly used not only for text categorization tasks but also for any other pattern classification tasks [11].

Neural Networks may be considered as an approach to text categorization, and among them, the MLP (Multiple Layers Perceptron) with back propagation is the most popular model. The neural network model was initially created in 1986 by Mcelland and Rumelhart, and it was intended to for performing tasks of pattern classification and nonlinear regressions as a supervised learning algorithm [12]. It was initially applied to text categorization in 1995 by Wiener [13]. Its performance was validated by comparing it with KNN in his master thesis on the test bed, Reuter21578. [13]. Even if the neural network classifies documents more accurately, it takes very much time for learning training documents.

The string kernel was proposed as the solution to the two main problems which is inherent in encoding documents into numerical vectors. It was initially proposed by Lodhi et al in 2002 as the kernel function of SVM [15]. The string kernel receives two raw texts as its inputs and computes their syntactical similarity between them. Since documents don't need to be encoded into numerical vectors, the two main problems are naturally avoided. However, it costed very time for computing the similarity and failed to improve the performance of text categorization.

3   String Vector

This section is concerned with the general aspect of string vectors. A string vector refers to an ordered set of strings. In the context of a natural language, a string indicates a word or a vocabulary. In other words, words or vocabularies are given as elements of string vectors. Therefore, this section describes in detail string vectors as the alternative representation of documents to numerical vectors.

A string vector is defined as a set of words which is ordered and has its fixed size. It is denoted by $[s_1, s_2, ..., s_d]$ where s_i denotes a string, and there are d elements. When representing documents into string vectors, their sizes are fixed with d, and it is called the dimension of string vectors. Since the elements are ordered in each string vector, two string vectors with their identical elements but different orders are treated as different ones. The reason is that each position of an element has its own different feature.

4   Neural Text Categorizer

The architecture of the NTC is illustrated in figure 1. The input layer receives an input vector given as a string vector and the number of nodes in the layer is consistent with the dimension of the string vector. The output layer generates categorical scores which indicates the likelihood of the given input vector to each category, and the number of nodes in the layer is consistent with the number of the predefined categories or classes. The learning layer decides the weights between the input and output layers differently depending on the given input vector, and the number of nodes in the layer is also consistent with the number of categories or classes. Therefore, with respect to its architecture, the NTC has the three layers as shown in figure 1.

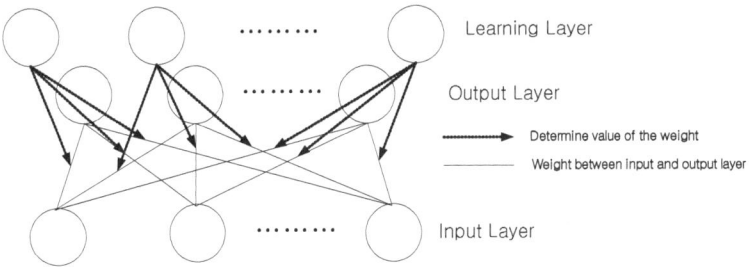

Fig. 1. Overall Architecture of the NTC

In context of the learning process, the first step of the NTC is to initialize the weights between the input and output layers. Let's assume that the NTC is applied to text categorization without decomposing the task into binary classification tasks. A set of the training string vectors is partitioned category by category. Each learning node has its own table which consists of words and their weights. Frequencies of elements of string vectors within each category assigned in the table as the initial weights. Therefore, the initial step is to set up the tables in learning nodes.

The learning process of the NTC refers to the process of optimizing the weights in the tables of the learning nodes. Each training example is classified by summing the initial weights and selecting the category corresponding to the maximal sum. If the training example is classified correctly, the weights are not updated. Otherwise, the weights are incremented toward the target category and those are decremented toward the classified category. The optimized weights are generated as the output of this process.

In the NTC, each example is classified by summing the optimized weights, whether it is a training or unseen example. Each output node generates the summation of weights connected to itself from the input nodes as its categorical score. The weights are decided by referring the table which is owned by its corresponding learning node. The category corresponding to the output node which generate its maximum categorical score is decided as the category of the given example. Therefore, the output of this process is one of the predefined categories, assuming that the NTC is applied to text categorization without the decomposition.

5   Empirical Results

The partition of the test bed, NewsPage.com into the training and test set is illustrated in table 1. The test bed is given as a small collection of news articles for entering the first evaluation, and its source is the web site, www.newspage.com[1]. The collection was built by copying and pasting news articles individually as the

---

[1] The test bed was built in 2002, but the web site was recently closed.

plain text files. In the test bed, the five categories and the 1,200 news articles are available. The collection of news articles is partitioned into the training and test set by the ratio 7:3, as shown in table 1.

Table 1. Training and Test Set in the Test Bed: NewsPage.com

Category Name	Training Set	Test Set	Total
Business	280	120	400
Health	140	60	200
Law	70	30	100
Internet	210	90	300
Internet	140	60	200
Total	840	360	1200

The configurations of the involved approaches are illustrated in table 2. The parameters of the SVM and the KNN, the capacity and the number of nearest neighbors, are set as four and three, respectively but the NB has no parameter. The parameters of the NNBP such as the number of hidden nodes and the learning rate are arbitrary set as shown in table 2. News articles are eneocded into 500 dimensional numerical vectors and 50 dimensional string vectors. Therefore, the configurations of the involved approaches are set as shown in table 2.

Table 2. The Configurations of Participating Approaches

Approaches to Text Categorization	Parameter Settings
SVM	Capacity = 4.0
KNN	#nearest number = 3
Naive Bayes	N/A
NNBP with Back Propagation	Hidden Layer: 10 hidden nodes Learning rate: 0.3 #Training Epochs: 1000
NTC	Learning rate: 0.3 #Training Epochs: 100

The results from this set of experiments are presented in figure 2. The y-axis in figure 00 indicates the accuracy as the rate of the correctly classified test examples to the entire test examples. As shown in figure 00, the four approaches participate in the comparison: KNN (K Nearest Neighbor), NB (Na?ve Bayes), the MLP (Multilayer Perceptron) with the back propagation, and the NTC as the proposed approach. As shown in figure 00, the NTC has its best performance among the approaches. Therefore, this set of experiments shows that the NTC is feasible in the text categorization even without the decomposition.

As shown in figure 2, the NTC has its better performance than any other approach. The NTC was previously compared with the approaches in the text

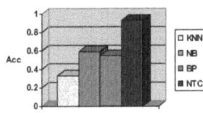

Fig. 2. The Results of This Set of Experiments

categorization with the decomposition. In the previous experiment, the NTC had its better performance than NB, and KNN, but it was comparable with the MLP with the back propagation. All of four approaches have worse performance than the case with the decomposition; there is the trade-off between the decomposition and the classification performance. In this set of experiments, the benefit of the NTC is that it has the least performance loss among the four approaches.

6   Conclusion

In this research, we discovered the feasible approach to the text categorization without the decomposition. By comparing the proposed approach with other approaches, we validated the feasibility in section 5. Previously, the decomposition was regarded as the mandatory course for using any machine learning algorithm for the text categorization. It was shown that without the decomposition, it the NTC has its significantly better performance. Before developing a light version of the text categorization system, the proposed approach needs to be applied to a large scaled text categorization where more categories are predefined as a remaining task.

References

1. Sebastiani, F.: Machine Learning in Automated Text Categorization. ACM Computing Survey 34(1), 1–47 (2002)
2. Cover, T.M., Hart, P.E.: Nearest Neighbor Pattern Classification. IEEE Transaction on Information Theory 13, 21–27 (1967)
3. Massand, B., Linoff, G., Waltz, D.: Classifying News Stories using Memory based Reasoning. In: The Proceedings of 15th ACM International Conference on Research and Development in Information Retrieval, pp. 59–65 (1992)
4. Yang, Y.: An evaluation of statistical approaches to text categorization. Information Retrieval 1(1-2), 67–88 (1999)
5. Kononenko, I.: ID3, sequential Bayes, naive Bayes and Bayesian neural networks. In: The Proceedings of 4th European Working Session on Learning, Montpellier, pp. 91–98 (1989)
6. Mitchell, T.M.: Machine Learning. McGraw-Hill, New York (1997)
7. Mladenic, D., Grobelink, M.: Feature Selection for unbalanced class distribution and Naive Bayes. In: The Proceedings of International Conference on Machine Learning, pp. 256–267 (1999)
8. Hearst, M.: Support Vector Machines. IEEE Intelligent Systems 13(4), 18–28 (1998)

9. Joachims, T.: Text Categorization with Support Vector Machines: Learning with many Relevant Features. In: The Proceedings of 10th European Conference on Machine Learning, pp. 143–151 (1998)
10. Drucker, H., Wu, D., Vapnik, V.N.: Support Vector Machines for Spam Categorization. IEEE Transaction on Neural Networks 10(5), 1048–1054 (1999)
11. Cristianini, N., Shawe-Taylor, J.: Support Vector Machines and Other Kernel-based Learning Methods. Cambridge University Press, Cambridge (2000)
12. McClelland, J., Rumelhart, D.: Parallel Distributed Processing, vol. 1,2. MIT Press, Cambridge (1986)
13. Wiener, E.D.: A Neural Network Approach to Topic Spotting in Text, The Thesis of Master of University of Colorado (1995)
14. Ruiz, M.E., Srinivasan, P.: Hierarchical Text Categorization Using Neural Networks. Information Retrieval 5(1), 87–118 (2002)
15. Lodhi, H., Saunders, C., Shawe-Taylor, J., Cristianini, N., Watkins, C.: Text Classification with String Kernels. Journal of Machine Learning Research 2(2), 419–444 (2002)
16. Estabrooks, A., Jo, T., Japkowicz, N.: A Multiple Resampling Method for Learning from Imbalanced Data Sets. Computational Intelligence 28(1), 18–36 (2004)

A Comparative Analysis of XML Schema Languages*

Muhammad Shahid Ansari, Noman Zahid, and Kyung-Goo Doh[**]

PLASSE Laboratory, Department of Computer Science and Engineering
Hanyang University, Ansan, South Korea

Abstract. XML (Extensible markup language) is playing progressively more important role in the exchange of wide variety of data on the Web and in many other applications. XML is not only used for data exchange, but also has it changed the concept of data formats, and that is the reason why it has become a standard and powerful data structure on the Internet. XML has simple syntax and can thus compose well-formed documents by following simple rules. To represent and validate these data structures and to check a well formed XML document, several validation languages are used. These validation languages are known as Schema languages. This paper presents a comparative analysis of four schema languages focusing on how these four languages are used, what data types are used in them, and how they are syntactically different from one another. The languages discussed in this paper are: DTD, XML Schema, RelaxNG, and DSD2. The benefit of the comparative analysis is to give the basic knowledge about two newly introduced schema languages (DSD2 and RelaxNG) against well-known and widely used languages (DTD and XML Schema). This comparison gives some basic knowledge about selected schema languages, illustrates the difference in their syntax, and demonstrates where to use which schema language.

Keywords: DSD2, DTD, RelaxNG, Schema, Schema processor, XML, XML document, XML Schema.

1 Introduction

This paper is about comparative analysis of the four most widely used schema languages for XML. This comparative analysis is focused on the pros and cons of these languages and the features they support. This analysis is of great importance in a situation where the best schema language has to be selected from the available schema languages.

The approach used for our analysis is divided into two parts. First, the features of these four schema languages are compared, and then the strengths and weaknesses of each language are pointed out. A feature-based comparison allows us to go into the

---

[*] This work was supported by the Engineering Research Center of Excellence Program of Korea Ministry of Education, Science and Technology (MEST) / Korea Science and Engineering Foundation(KOSEF), R11-2008-007-01003-0.
[**] Corresponding author.

D. Ślęzak et al. (Eds.): DTA 2009, CCIS 64, pp. 41–48, 2009.

details of the schema languages at the smallest possible level. An instance of XML document is devised in order to show the characteristics of each schema, which will be helpful in studying the strengths and weaknesses of each language. The conclusion is drawn based upon this observation.

Normally XML document is referred as an XML tree or instance document. A schema for XML provides a formal definition of the structure of XML documents. We have chosen four schema languages, all supported by major standardization organizations: DTD, XML Schema, RelaxNG, and DSD2.

The rest of the paper is structured as follows. In Section 2, backgrounds about XML and schema languages are given. Comparison among schema languages is shown in Section 3. Last but not the least Section 4 is comprised of conclusion.

2 Background

XML stands for *eXtensible Markup Language*. It is a markup language much like HTML and is designed to carry data, not to display data. XML tags are not predefined. You must define your own tags. It is designed to be self-descriptive. XML is a W3C Recommendation. XML is created to structure, store, and transport information [1]. An XML document can be represented in a tree structure that is the reason why it is also termed as XML tree. Nodes in XML tree represent elements, texts, attributes, and other information that an XML document can hold. XML is now as important for the Web as HTML was to the foundation of the Web. XML is the most common tool for data transmissions today, and is becoming more and more popular in the area of storing and describing information.

2.1 Schema Languages

A program in an XML schema language is a template to define transformations to XML instance documents. In fact, schema is a formal syntax definition of an XML instance document. These schemas take instance documents as input and produce a validation report, which indicates whether the documents are valid or not.

Every schema language is designed in the form of *schema processor*, so that it can be used for the validation of an XML document. Mechanism of validating an XML document using schema language is shown in Figure 1.

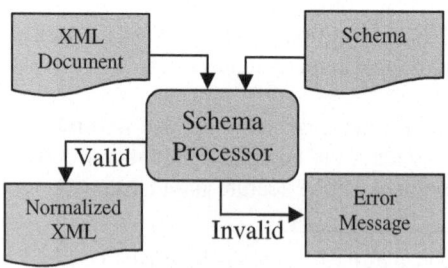

Fig. 1. Schema Processing Model

A schema language must satisfy three criteria to be useful [3]:

- The language is expressible enough so that most syntactic requirements can be formalized in the language.
- The language is implementable as an efficient schema processor.
- The language is comprehensible by non-experts.

Some of schema languages support *Post-Schema-Validation Infoset* (PSVI). Validating an XML document using the schema produces a PSVI, where any node of the XML document has assigned the type as defined in XML Schema.

2.1.1 DTD

A Document Type Definition (DTD) defines the legal building blocks of an XML document. It defines the document structure with a list of legal elements and attributes [2]. This document type definition could be given inside an XML document, or external reference could be written in that XML document. Your application can use a standard DTD to verify the data that you receive from the different outside sources as well as you can verify your own data. Unfortunately, there is no support of namespaces in DTD. There are a number of problems in DTD that limits its usage practically, but regardless of all those problems and new schema languages in market, DTD is still widely used.

2.1.2 XML Schema

Shortly after the approval of XML 1.0 and DTD, by W3C, the development of next generation of schema languages was started to cover up the problems with DTD. An XML schema describes the structure of an XML document. The XML Schema language is also referred to as XML Schema Definition (XSD). The purpose of an XML Schema is to define the legal building blocks of an XML document, just like a DTD [7]. XML Schema gives the description of elements and attributes appearing in a document. XML Schema is a strongly typed schema language often considered complex because of the number of features it supports. Being one of the most powerful schema languages, XML Schemas are now used in most Web applications as a replacement for DTDs.

2.1.3 DSD2

The DSD2 (Document Structure Description 2.0) language is a successor to the DSD 1.0 language developed in cooperation by the University of Aarhus and AT&T [3]. It is designed to be a simple and easily understandable language. DSD2 has good support for the structure of instance documents, data types, and identity-constraints. This is reflected in the fact that the specification [4] is only 15 pages excluding examples, and it is 100% self-describing. As compared with a few other schema languages, DSD2 is small, supports Boolean logic and regular expressions, and has more expressive power than other schema languages.

2.1.4 RelaxNG

The RelaxNG schema language has been developed within the Organization for the Advancement of Structured Information Standards (OASIS) and is being standardized by ISO. This language has been designed with the same fundamental goals as DSD2; simplicity and expressiveness. However, the design is quite different. DSD2 is based on rules while RelaxNG is based on grammars [3].

RelaxNG is simple, is easy to learn, and supports XML namespaces. Even though RelaxNG is considered technically superior to DTD and XML Schema, but it is not widely used by software developers and vendors yet.

3   Comparing Schema Languages

Table 1 shows the feature comparison among four examined schema languages. This comparison is based upon the basic features supported by languages. In Table 1, horizontally shown entities are the schema languages, while features to be compared in between them are shown vertically.

Table 1. Comparison of Language Features

	DTD	*XML Schema*	*DSD2*	*RelaxNG*
PSVI	Yes	Yes	No	No
Structures	Yes	Yes	Yes	Yes
Data Types	Weak	Yes	Yes	No (Plug-in is available)
Rules	No	No	Yes	No
Namespaces	No	Structures	Yes	Yes
XML Syntax	No	Yes	Yes	Yes
Other Syntaxes	No	No	No	Yes
Grammar Based	Yes	Yes	No	Yes
Recommendation	W3C	W3C	BRICS	OASIS RelaxNG Committee Specification

Table 2 compares Elements, and Table 3 compares Attributes in these languages.

Table 2. Comparison of Elements

	DTD	*XML Schema*	*DSD2*	*RelaxNG*
Element Conditional Definition	No	No	Yes	Yes
Uniqueness in Element	No	Yes	Yes	No

Table 3. Comparison of Attributes

	DTD	*XML Schema*	*DSD2*	*RelaxNG*
Required	Yes	Yes	Yes	Yes
Optional	Yes	Yes	Yes	Yes
Attribute Conditional Definition	No	No	Yes	Yes
Uniqueness in Attributes	Yes	Yes	Yes	No

Following is the example of an XML document:

```
<?xml version="1.0"?>
<inbox eid="noman">
    <mail id="a001">
        <from>Shahid</from>
        <subject>Hello World</subject>
        <body> How are you buddy? </body>
        <attachment>
            <fpath>/abc.doc</fpath>
            <fpath>/topic.doc</fpath>
        </attachment>
    </mail>
    <mail id="a002">
        <from>Subi</from>
        <subject>Congrats</subject>
        <body> Congratulations. </body>
    </mail>
</inbox>
```

Rest of this section will show the example codes of XML schema languages for above given XML document.

DTD: Schema definition is only one of the features of DTDs. DTD does not use XML syntax so it is not possible to write a DTD description of a DTD language itself. DTD for example XML document is as follow:

```
<?xml version="1.0" encoding="UTF-8"?>

<!ELEMENT inbox ( mail* ) >
<!ATTLIST inbox eid NMTOKEN #REQUIRED >

<!ELEMENT mail ( from, subject, body, attachment? ) >
<!ATTLIST mail id NMTOKEN #REQUIRED >
<!ELEMENT from ( #PCDATA ) >
<!ELEMENT subject ( #PCDATA ) >
<!ELEMENT body ( #PCDATA ) >

<!ELEMENT attachment ( fpath+ ) >
<!ELEMENT fpath ( #PCDATA ) >
```

XML Schema: Once you write an XML Schema, it can be extended for further additions. XML Schemas are wealthier and more influential than DTDs, and written in XML format. XML Schema for example XML document is as follow:

```
<?xml version="1.0" encoding="UTF-8"?>
<xs:schema xmlns:xs="http://www.w3.org/2001/XMLSchema"
           elementFormDefault="qualified">
    <xs:element name="inbox">
        <xs:complexType>
            <xs:sequence>
                <xs:element maxOccurs="unbounded" ref="mail"/>
            </xs:sequence>
        <xs:attribute name="eid" use="required" type="xs:NCName"/>
        </xs:complexType>
    </xs:element>
    <xs:element name="mail">
        <xs:complexType>
            <xs:sequence>
                <xs:element ref="from"/>
```

```
                <xs:element ref="subject"/>
                <xs:element ref="body"/>
                <xs:element minOccurs="0" ref="attachment"/>
            </xs:sequence>
            <xs:attribute name="id" use="required" type="xs:NCName"/>
        </xs:complexType>
    </xs:element>
    <xs:element name="from" type="xs:NCName"/>
    <xs:element name="subject" type="xs:string"/>
    <xs:element name="body" type="xs:string"/>
    <xs:element name="attachment">
        <xs:complexType>
            <xs:sequence>
                <xs:element maxOccurs="unbounded" ref="fpath"/>
            </xs:sequence>
        </xs:complexType>
    </xs:element>
    <xs:element name="fpath" type="xs:string"/>
</xs:schema>
```

DSD2: DSD2 is based on rules and is self-descriptive. DSD2 itself is described in DSD2[1]. DSD2 for example XML document is as follow:

```
<dsd xmlns="http://www.brics.dk/DSD/2.0" xmlns:e="example"
    xmlns:c="http://www.brics.dk/DSD /character-classes"
    root="e:inbox">

<import href="http://www.brics.dk/DSD /character classes.dsd"/>

<if><element name ="e:inbox"/>
    <declare> <required>
        <attribute name="eid">
                <stringtype ref="e:NMTOKEN"/>
            </attribute>
        </required> <contents>
        <sequence>
            <repeat> <element name="e:mail"/> </repeat>
        </sequence>
        </contents> </declare>
         <unique> <and>
            <element name="e:mail"/>
            <attribute name="id"/>
        </and> <attributefield name="id"/>  </unique>
    </if>
      .
      .
      .

    (Being very long this code is cut short. Rest of the code will
proceed same way for remaining tags of the example XML document.)
      .

      .
    </dsd>
```

RelaxNG: Unlike DSD2, RelaxNG that is based on grammar has two types of syntax: one is XML syntax and the other is non-XML syntax. Non-XML syntax is compact, easier to read and straightforward to understand. For the XML-syntax of RelaxNG, a RelaxNG itself is described in RelaxNG[2]. XML syntax RelaxNG and non-XML RelaxNG for example XML document are respectively as follow:

---

[1] http://www.brics.dk/DSD/dsd2.dsd
[2] http://www.relaxng.org/relaxng.rng

XML-syntax RelaxNG:

```
<element xmlns="http://relaxng.org/ns/structure/1.0" datatypeLibrary
= "http://relaxng.org/ns/compatibility/datatypes/1.0" name="inbox">
     <attribute name = "eid" >
          <data type= "ID"/>
     </attribute>

     <oneOrMore>
          <element name = "mail">
          <attribute name = "id" >
               <data type= "ID"/>
          </attribute>
               <element name = "from"> <text/> </element>
               <element name = "subject"> <text/> </element>
                <element name = "body"> <text/> </element>

               <optional>
                    <element name = "attachment">
                         <oneOrMore>
                              <element name = "fpath"> <text/> </element>
                         </oneOrMore>
                    </element>
               </optional>
          </element>
     </oneOrMore>
</element>
```

Non-XML syntax RelaxNG:

```
datatypes d = "http://relaxng.org/ns/compatibility/datatypes/1.0"

element inbox {
     attribute eid { d:ID },
     element mail {
          attribute id { d:ID },
          element from { text },
          element subject { text },
          element body { text },
          element attachment {
               element fpath { text }+
          }?
     }+
}
```

4 Conclusion

By comparing four schema languages using their features and sample schemas codes, it can be concluded that there is no perfect XML Schema language. DTD, XML Schema, DSD2 and RelaxNG, all have some advantages and disadvantages. The selection of schema language depends on the scenario: if you need restrict data-type checking then RelaxNG could be the best, just for structure validation, DTD might be better, and for applying rules over your XML instance document, DSD2 would be the best choice. There is no doubt that currently most influential, powerful and widely used schema language is XML Schema. DSD2 and RelaxNG are not widely used, comparing with DTD and XML Schema. Unlike DTD and XML Schema, DSD2 and

RelaxNG are not even supported by daily used browsers because they were primarily built for academic purpose. With the help of separate tools, XML can be validated for DSD2 and RelaxNG. All schema languages have their own strengths and weaknesses, and what you have to do is to choose the right one for your application.

References

1. http://www.w3schools.com/XML/xml_whatis.asp
2. http://www.w3schools.com/DTD/dtd_intro.asp
3. Møller, A., Schwartzbach, M.I.: An Introduction to XML and Web Technologies. Addison-Wesley, Reading (2006)
4. Møller, A.: Document Structure Description 2.0, BRICS, Department of Computer Science, University of Aarhus, Notes Series NS-02-7 (December 2002), http://www.brics.dk/DSD/
5. http://www.brics.dk/DSD/dsd2.html
6. http://www.oasis-open.org/committees/relax-ng/spec-20011203.html
7. http://www.w3schools.com/Schema/default.asp

Mining Approximate Frequent Itemsets over Data Streams Using Window Sliding Techniques*

Younghee Kim, Eunkyoung Park, and Ungmo Kim

School of Information and Communication Engineering, Sungkyunkwan University,
300 Chunchun-dong, Suwon, Gyeonggi-Do, 440-746, Korea
younghees@gmail.com, wjbest527@gmail.com, umkim@ece.skku.ac.kr

Abstract. Frequent itemset mining is a core data mining operation and has been extensively studied in a broad range of application. The frequent data stream itemset mining is to find an approximate set of frequent itemsets in transaction with respect to a given support threshold. In this paper, we consider the problem of approximate that frequency counts for space efficient computation over data stream sliding windows. Approximate frequent itemsets mining algorithms use a user-specified error parameter, ε, to obtain an extra set of itemsets that are potential to become frequent later. Hence, we developed an algorithm based on the Chernoff bound for finding frequent itemsets over data stream sliding window. We present an improved algorithm MAFIM (a *maximal approximate frequent itemsets mining*) for frequent itemsets mining based on approximate counting using previous saved maximal frequent itemsets. The proposed algorithm gave a guarantee of the output quality and also a bound on the memory usage.

Keywords: Data Stream, Maximal approximate frequent itemsets, Potential frequent itemsets, Chernoff bound.

1 Introduction

Mining data streams for knowledge discovery is important to many applications, such as fraud detection, intrusion detection, trend learning, etc. Many previous studies contributed to the efficient mining of frequent itemsets over data streams. Manku and Motwani (2002) proposed the Lossy Counting algorithm to evaluate the approximate support count of a frequent itemset in data streams based on the landmark model [1]. Giannella et al. (2003) proposed techniques for computing and maintaining all the frequent itemsets in data streams. The proposed FP-stream algorithm is to find the set of frequent itemsets at multiple time granularities by a novel titled-time windows technique [2]. Chang and Lee (2003) developed a damped window based algorithm, called *estDec*, for mining frequent itemsets over streaming data in which each transaction has a weight decreasing with age [3]. Lee et al (2004) proposed an efficient

---

* This work was supported by the Korea Science and Engineering Foundation (KOSEF) grant funded by the Korea government (MEST) (No.2009-0075771).

D. Ślęzak et al. (Eds.): DTA 2009, CCIS 64, pp. 49–56, 2009.

method for incremental mining, called SWF, to deal with the candidate itemset generation by partitioning a transaction database into several partition [4]. Yu et al. proposed a false-negative approach. This method focuses on the entire history of a data stream and does not distinguish recent itemsets from old ones [5]. In this paper, we resent an improved algorithm MAFIM (a *maximal approximate frequent itemsets mining*) for frequent itemsets mining based on approximate counting using previous saved maximal frequent itemsets. Chernoff bound based algorithm is given, that guarantee of the output quality and also a bound on the memory usage. Therefore, our algorithm is controlled by two parameters ε and δ for error bounds and reliability. We can see a reasonable value for ε, accurate result, fast computation and low memory utilization can be achieved.

2 Preliminaries

Suppose there is a sequence of elements, $e_1, e_2, ..., e_i, ..., e_N$, in data stream and consider the first n $(n \ll N)$ observations as independent Bernoulli trails(coin flips) such that $Pr(head) = p$ and $Pr(tail) = 1 - p$ for a probability p. Let k be the number of heads in the n coin flips. Then, the expectation of k is np. Chernoff bound states, for any $\sigma > 0$.

$$\Pr[|k - np| \geq np\sigma] \leq e^{-\sigma^2/4} \leq 2e^{-np\sigma^2/2} \tag{1}$$

By substituting $\bar{k} = k/n$ and $\varepsilon = p\sigma$,

$$\Pr\left[|\bar{k} \times n - np| \geq n\varepsilon\right] \leq 2e^{-np\sigma^2/2}$$

$$= \Pr\left[|\bar{k} - p| \geq \varepsilon\right] \leq 2e^{-np\sigma^2/2} \leq 2e^{-\frac{n\varepsilon^2/p}{2}} \leq 2e^{-n\varepsilon^2/2p} \quad (\text{by } \sigma = \varepsilon/p) \tag{2}$$

Let $\delta = 2e^{-n\varepsilon^2/2p}$. We obtain the following equation, where δ is the reliability parameter.

$$\frac{-n\varepsilon^2}{2p} = \frac{\log_2 \delta}{\log_2 2e}, \quad \varepsilon^2 = \frac{2p \times \log_{2 \times e} \delta}{-n}, \quad \varepsilon = \sqrt{\frac{2p \times \log_{2 \times e} \delta}{n}} = \sqrt{\frac{2p \times \ln(\frac{2}{\delta})}{n}} \tag{3}$$

By $s = p$, the minimum support s as the probability p.

$$\varepsilon = \sqrt{\frac{2s \times \ln(\frac{2}{\delta})}{n}} \tag{4}$$

Then, from equation (3), we can produce $\bar{k}$ satisfying:

$$\Pr[s - \varepsilon \leq \bar{k} \leq s + \varepsilon] \geq 1 - \delta \tag{5}$$

In other word, for an itemset X, true support of X is within $\pm\varepsilon$ of s with reliability $1 - \delta$. As shown Fig 1, we can see that, our bound does not rely on the user-specified σ, but on a chernoff bound ε which decreases while the number of observations n increases.

If we can set a reasonable value for ε, accurate result, fast computation and low memory utilization can be achieved. As a result, we believe that our method, promising to solve our algorithm.

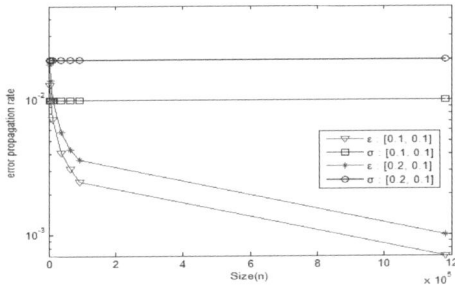

Fig. 1. The rate of error propagation within the range $[s, \delta]$

3   MAFIM Algorithm

In this section, we develop an algorithm based on the chernoff bound for mining frequent itemsets, called the MAFIM. The proposed method is maintained some approximate frequent itemsets that is computed over a sliding window of the data stream.

3.1   Basic Concept of MAFIM Method

Assume that the current data stream, $DS_m = \{e_1, e_2, ...e_m\}$ when the current size of the stream is m and the current size of window is N. In addition, when the size of a sliding window is denoted by W, the current window $DS_{(W,N)} = \{e_{m-N+1}, e_{m-N+2}, ..., e_m\}$ in the current data stream DS_m is defined by the set of transactions that are most recently generated. A data stream is decomposed into a sequence of block, which are assigned with |W| of length. Let the block numbered i be denoted as B_i. The number of transactions in B_i is denoted as $|B_i|$ and for each B_i, the current window that consists of the |W| consecutive blocks from $B_{i-|W|+1}$. The main procedure of algorithm MAFIM is outlined below. As shown Fig 2, the MAFIM algorithm is conducted the impacts of a large number itemsets in the range of $[\sigma - \varepsilon, \sigma + \varepsilon]$ on frequent itemsets mining over active window. The frequent itemsets generation is described as follows. First of all, our algorithm read every transaction from each current block in current

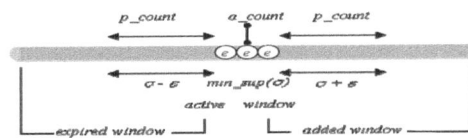

Fig. 2. The approximate support of frequent itemsets in active window

window. Then, we keep the potential frequent itemsets in the active table with respect to σ in each block B. For the first block B, after finding the potential frequent itemsets, it is set to the support of each itemset. We obtain ε based on the chernoff bound and both are initialized potential frequent itemset support p_count, and use a_count for approximate count in active window. When B_i arrives, where $1 \le i \le |W|$, three conditions are executed one by one from 4 to 14. The proposed algorithm can be defined as a sequence of two operations: insertion of a new stream element into the window when it arrives, and deletion of the oldest element in the window. Basically, as shown Fig 3, the process of MAFIM algorithm can be generalized into three steps. The *window initialization phase* is activated that after $m \ge N (= W)$ elements of the stream. In this phase, each element of the new incoming transaction is approximately counted. In current window, $DS_{(W,N)}$ allows $\varepsilon \cdot e_i / N$ approximate count. In *frequent itemsets generation phase*, by given the user support threshold σ ($0<\sigma<1$) and chernoff bound ε ($0< \varepsilon < \sigma$), the frequent itemsets in a current window of size N can be divided into three patterns. The *potential frequent itemsets* may become frequent later. We are only interested in *frequent itemsets*, and *infrequent itemsets* will discard because the number of infrequent itemsets is really large over data stream. Hence, the error will be no more than ε because of the loss of support from infrequent itemsets. It is guarantees that no false negative. In *updating window phase*, the arrival of a new block also triggers in current window, which are differently executed in three cases: New itemsets insertion, Old itemsets update and Itemsets discounting.

Algorithm 1. MAFIM algorithm

Input: DS_{db} (a transaction data stream), $|W|$ (window size), ε (an estimated the chernoff bound)

　　　σ (a user defined minimum support threshold in the range of [0,1]), δ (probability)

Output: a set of frequent itemsets $\widetilde{Fi}_{max}$

---

1. **begin**	13. Generate the frequent *k-itemsets*;
2. $a\_count = 0$; $p\_count = 0$; $W = $ NULL;	14. **end for**
3. **for each** new transaction T_i in W **do**	15. **for each** new block **do**
4. **if** $W = $ FULL **then**	/* a new incoming block */
5. **for all** itemset e in T_i **do**	16. $a_count = $ current window(a_count)
6. $a\_count = a\_count + 1$;	+ new block(a_count)
7. **if** ($a\_count < \sigma$)	17 **if** ($a\_count < \sigma$)
8. delete $a\_count.table(e)$	18 delete $a\_count.table(e)$
9. add $p\_count.table(e)$	19 add $p\_count.table(e)$
10. **end for**	20. **end for**
11. output $\widetilde{Fi}_{(k=1)}$	21.delete the oldest block ($a\_count(e)$ &
/* {frequent 1-itemsets} */	$p\_count(e)$);
12. **for** ($k=2$; $\widetilde{Fi}_{(k-1)} \ne$ NULL; $k++$) **do**	22. Output $\widetilde{Fi}_{max}(a\_count \ge \varepsilon N)$

Fig. 3. MAFIM algorithm

3.2 Mining of Potential Frequent Itemsets within a Current Sliding Window

In this section, we give examples to explain the approach of MAFIM algorithm. Assume that the current window for mining is W_1, the size of sliding windows is 3 and the size of each buffer is 10. Let the user defined minimum support threshold σ is

0.5 and the estimated chernoff bound ε is 0.1. In initialization step, *a_count* in the active table and the *p_count* in potential table are set to be 0. The example is shown in Table 1. By Step 1, the algorithm MAFIM, all the transaction data in the block B_1 is scanned once first. The *a_count* value for each item is evaluated. For example, in block B_1, the *a_count* of item "a" is 6. If the *a_count* at the block B_1 is less than the minimum support threshold, these items are removed. Then, we keep support count (*p_count*) in potential table. Thus, it would be considered as a potential frequent itemsets in the next step.

Table 1. The transaction data in current window W_1

TID	T_1	T_2	T_3	T_4	T_5	T_6	T_7	T_8	T_9	T_{10}
B_1	abcd	abcde	bc	bcde	abcde	bcd	ad	abcd	bde	abc
B_2	bcd	abe	abc	bce	abce	bce	ade	abc	be	abcde
B_3	bc	ace	abc	abe	bce	ace	abd	abc	be	abce
B_4	abe	ace	abce	ade	bce	ace	abe	abe	bce	Abce

Next, candidate 2-*itemsets* are generated using these frequent 1-*itemsets*. In this step, MAFIM algorithm uses an apriori property to generate the set of candidate itemsets with k items from frequent itemsets k-1 in the step 1. The candidate generate process is stopped until no new candidates with k+1 items are generated. For example, our algorithm generates five candidate 2-*itemsets*, {ab}, {ac}, {bc}, {bd} and {cd}, by combining frequent 1-*itemsets* in step 1. We removed in active table becuase 2-*itemset* {ad} is an infrequent itemset. Similarity, 3-*itemsets* and 4-*itemsets* generate according to apriori property and the process is stopped. Hence, there are the frequent itemsets, {{a}, {b}, {c}, {d}, {ab}, {ac}, {bc}, {bd}, {cd}, {abc}, {bcd}}, generated by MAFIM algorithm in block B_1. Similarly, frequent itemsets generation in block 2, candidate 2-itemsets are generated using frequent 1-*itemsets*. Therefore, *p_count* of candidate itemsets {ac}, {ae}, {ce}, {abc}, {abe}, and {bce} set the potential table. Then we show all the frequent itemsets: {{a}, {b}, {c}, {e}, {ab}, {bc}, {be}}. Next, the *a_count* of {d} in block 3 is 1. Because the *a_count* of {d} is less than the minimum support threshold, it is removed from active table. After, we evaluate the *p_count* of {d} according to *p_count* of {d} in previous step. For example, the *p_count* of {d} is the summation of the previous support values, i.e., 3+1 = 4. To find frequent itemsets on a data stream, we cannot keep the support of all itemsets, because of time and space constraints. On the other hand, we may prevent lose information of the potential frequent itemsets over a sliding window. In block 3, all the frequent itemsets generated are {{a}, {b}, {c}, {e}, {ab}, {ac}, {bc}}. *The block merge phase* is activated after the current sliding window W_1 becomes full. All the maximal frequent itemsets of active and potential table for each block are evaluated. For example, in block 1, the maximal frequent itemsets are {{abc}, {bcd}} and the potential frequent itemsets is {abcd}. *The frequent itemsets generation phase* is performed after the maximal frequent itemsets in each block is generated. In block B_1, 1-*itemset* {a}, {b}, {c}, and {d} is evaluated by {abc} and {bcd}. For example, the max value of itemset {b} is 6, i.e., the *a_count* of {abc}=5 and the *a_count* of {bcd}=6. All the frequent itemsets can be evaluated in a similar way using apriori property. Finally, the final frequent itemsets is evaluated by $[B_1 + B_2 + B_3]/|W|$. Then the frequent itemsets in current window W_1 is {{a}, {b}, {c}, {ab}, {bc}}. Similarly, the potential frequent itemsets are generated. Assume the estimated chernoff bound ε is 0.1, according to the definition of $(\sigma - \varepsilon) \times N$, for example $(0.5-0.4)\times 10$ is 4. Based on the above discussion, we design an improved method to maintain whether the potential frequent itemsets is keep or not.

3.3 Mining Frequent Itemsets Insert and Delete Phase

A new block is inserted into the window, as the oldest block (expired block) is deleted from the current window. In data stream, all the transaction data in each window are scanned once first. First, the support count of 1-*itemset* for a new block is evaluated. As shown Table 1, in block B_4, the a_count of each itemsets is {{a}: 8, {b}: 7, {c}: 6, {d}: 2, {e}: 9}. It is evaluated by the summation of the a_count of {d} in a inserted block and p_count of the potential table, because the a_count of itemsets {d} is less than the minimum threshold. The p_count of {d} is $2 + 4 = 6$. Next, $B_4 + W_1$ itemsets are generated using the maximal frequent itemsets in the current window. For example, the support of {b} is evaluated by $7(B_4) + \max$ [{ab}:5, {bc}:6] $(W_1) = 13$. Similarly, 2-*itemsets* {ab} is evaluated by $5(B_4) + 5(W_1) = 10$. Adding a new block to the current sliding window will be satisfied the one pass scan. From the above discussion and from the insert algorithm shown in Fig 4. Initially, after some insert process, old block are removed from the current window. After, an old block B_1 is deleted from the sliding window. In the following, we discuss the impact of deletion in detail. First, after an itemset in current window is deleted if its sum of a_count and p_count is less than $\varepsilon \times \sum B_i$. Since, the transactions in $B_{i-|W|}$ will the expired, the support counts of the itemsets kept by current window are discounted accordingly. For example, deletion may subtract the expired block B_1 from $B_4 + W_1$. Here, the a_count of the expired block B_1 are evaluated from the a_count of each itemset of maximal frequent itemsets after inserting. In this example, the a_count of itemset for $W_1 + B_4 - B_1$ is evaluated by the maximal frequent itemsets of {abc}:5 and {bcd}:6. Then the result of {a} is $8 = 13-5$. Thus all frequent itemsets are generated. {{a:8}, {b:7}, {c:6}, {e:13}, {ab:7}, {ac:6}, {ae:8}, {bc:6}, {be:7}, {ce:6}}. Next, the p_count of potential table of the determined itemsets set to be 0. Fig 4 gives the process of delete operation of an expired block.

3.4 Pruning Strategy

For deleting the infrequent itemsets and maintaining the potential frequent itemsets, we show an efficient pruning method. In the MAFIM algorithm, the constant value $\sigma \times |B_i|$ is the frequent threshold of itemsets, where σ is the user-defined minimum support threshold. When incrementally adding the new transactions, many infrequent itemsets are updated into the active table. To save the space of the active table, those infrequent itemsets should be deleted too when pruning the items. Suppose the size of the sliding window be N, and the operation of pruning itemsets be performed just after all transaction in the subsequent blocks has been processed.

Definition 1 (Pruning Itemsets). Let the minimum support threshold be σ and the error rate by an estimated the chernoff bound be ε. The itemsets $\widetilde{Fi}$ is pruned from the potential frequent itemsets only if it satisfies the following case.

$$Act = \left\{ \widetilde{Fi} \mid a\_count\left(\widetilde{Fi}\right) < \sigma \mid B_i \mid, \ \widetilde{Fi} \in active \ table \right\} \ // \ the \ \widetilde{Fi} \ delete \ from \ active \ table$$

$$Pot = \left\{ \widetilde{Fi} \mid p\_count\left(\widetilde{Fi}\right) < \left(\sigma \mid B_i \mid - \varepsilon \mid B_i \mid\right), \ \widetilde{Fi} \in potential \ table \right\} // \ the \ \widetilde{Fi} \ delete \ from \ potential \ table$$

(6)

In the formula 6, let *Act* be a set of itemsets that doesn't satisfy the true positive frequent itemsets. These itemsets are not frequent at this subsequent block in the current window. So we must delete these itemsets from active table. Then this method efficiently prunes the candidate frequent itemsets. Therefore we efficiently store the frequent itemsets and minimize the memory space. Let *Pot* be a set of itemsets that doesn't satisfy the false negative frequent itemsets. These itemsets are not frequent at this window and we can guarantee to minimize the missing true answers and dynamically control the level of the infrequent itemsets by our method.

Algorithm FIA_INSERT( )	**Algorithm FIA_DELETE ( )**

1. **if** ($N > \mid W \mid$ and (New block B_k = FULL)) **then**
 2. **for all** transaction $T_i(e)$ in new block **do**
 3. $a\_count = a\_count + 1$;
 4.   **if** $(a\_count < \sigma)$
 5.   delete $a\_count\_table(e)$;
 6. $p\_count = a\_count$;
 7. **end for**
 8. $a\_count(e)$=max[$a.count$(maximal frequent _
         itemsets($\widetilde{Fi}_{max}$))]
 9. $a\_count(e)$= current window(a_count) _
        + new block(a_count)
 10.   **if** $(a\_count(e) < \sigma)$ {
 11.   delete $a\_count.table(e)$
 12.   add $p\_count.table(e)$ }
 13. **for** $(k$=2; $\widetilde{Fi}_{(k-1)} \neq$ NULL; k++) **do**
 14.   Generate the frequent k-itemsets;
 15. **end for**
	1. **if** (a number of $B_i > \mid W \mid$) **then**
	 2. **for all** transaction $T_i(e)$ in expired block **do**
	 3. $a\_count(e)= a\_count$ (B_k)
	        + a_count(W_k-$\mid W \mid$)- a_count(B_k-$\mid W \mid$);
	 4. **if** $(a\_count < \sigma)$
	 5. delete $a\_count\_table(e)$;
	 6. $p\_count= a\_count + p\_count$;
	 7. **for** $(k$=2; $\widetilde{Fi}_{(k-1)} \neq$ NULL; k++) **do**
	 8. Generate the frequent k-itemsets;
	 9. **end for**
	 10. **if** $(e \in \widetilde{Fi}$) **then**
	 11. $p\_count(e) = 0$;
	 12. **end for**

Fig. 4. The insertion and deletion algorithm

4 Experimental Results

In this section, we show the experimental evaluation of the proposed MAFIM algorithms. We evaluate the performance of our MAFIM algorithm by varying the usage of the memory space. We also analyze the execution time. The simulation is implementation in Visual C++ and conducted in a machine with 3GHz CPU and 1GB memory. Fig 5(a) shows these numbers for the BMS-WebView-1 data set. BMS-WebView-1 is a relatively sparse data set. Fig 5(b) shows these numbers for the Mushroom data set under different minimum supports. The Mushroom data set is relatively dense data set. In the following experiments, the minimum support threshold σ vary from 0.1% to 1.0%, $\delta = 0.05$ in data sets. The size of the sliding window is 5K transactions. We compare our algorithm MAFIM with Lossy Counting and FDPM algorithm. As shown in Fig. 5, MAFIM significantly outperforms LC. From the figures we can see that the memory requirement of proposed algorithm in the frequent itemset mining process is less than that of LC and FDPM, because our

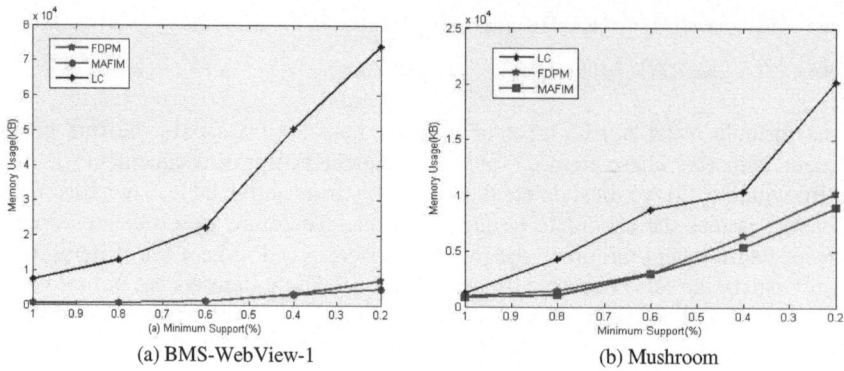

(a) BMS-WebView-1   (b) Mushroom

Fig. 5. Comparisons of memory usages in varying support σ

algorithm maintains only the maximal frequent itemsets of previous window. Therefore, the proposed our algorithm is a memory efficient method for frequent itemsets mining from data streams based on window sliding.

5 Conclusion

In this paper, we propose the MAFIM algorithm based on the Chernoff bound with a guarantee of the output quality and also a bound on the memory usage. Therefore, our algorithm is controlled by two parameters ε and δ for error bounds and reliability. We can set a reasonable value for ε, accurate result, fast computation and low memory utilization can be achieved. We evaluate the performance of our algorithm by varying the usage of the memory space.

References

1. Manku, G.S., Motwani, R.: Approximate Frequency Counts Over Data Streams. In: Proceedings of the 28th International Conference on Very Large Data Bases, pp. 346–357 (2002)
2. Giannella, C., Han, J., Pei, J., Yan, X., Yu, P.S.: Mining frequent patterns in data streams at multiple time granularities. In: Data Mining, Next Generation Challenges and Futures Directions, pp. 191–212. AAAI/MIT Press (2004)
3. Chang, J., Lee, W.: A Sliding Window Method for Finding Recently Frequent Itemsets over Online Data Streams. Journal of Information Science and Engineering 20(4) (July 2004)
4. Lee, C.H., Lin, C.R., Chen, M.S.: Sliding window filtering: An efficient method for incremental mining on a time-variant database. Information Systems 30, 227–244 (2005)
5. Yu, J.X., Chong, Z., Lu, H., Zhang, Z., Zhou, A.: False positive or false negative: mining frequent itemsets from high speed transactional data streams. In: Proc, VLDB, pp. 204–215 (2004)

Preserving Referential Integrity Constraints in XML Data Transformation*

Md. Sumon Shahriar and Jixue Liu

School of Computer and Information Science, Data and Web Engineering Lab
University of South Australia, SA-5095, Adelaide, Australia
shamy022@students.unisa.edu.au, Jixue.Liu@unisa.edu.au

Abstract. We study the transformation and preservation of XML referential integrity constraints in XML data transformation for integration purposes in this paper. In transformation and preservation, we consider XML inclusion dependency and XML foreign key. We show how XML referential constraints should be transformed and preserved using important transformation operations with sufficient conditions.

1 Introduction

Transformation of schema with its conforming data is an important task in many data intensive activities such as data integration[3,4] and data warehousing[1]. In recent days, XML[16] is widely used data representation and storage format over the web and hence the task of XML data transformation[2,15] for integration purposes is getting much importance to the database researchers and developers. In XML data transformation, a source schema with its conforming data is transformed to the target schema. An XML source schema is often defined with referential integrity constraints[5,6,7,8] such as XML inclusion dependency[9,10] and foreign key[11]. Thus, in transforming schema and its data, referential integrity constraints can also be transformed. There is also a need to see whether constraints are preserved[13,14] by the transformed data at the target schema.

We illustrate our research problems with following simple motivating examples.

Example 1. Consider the DTD D_1 in Fig.1(a) that describes the enrolled students(sid) of different courses(cid) and the students who has project in the department. Now consider the XML key[12] $\Bbbk_1(dept, \{enroll/sid\})$ on D_1. We say $dept$ is the $selector$ and $enroll/sid$ is the $field$ of key. We also say that the key is valid on D_1 because it follows the path of the DTD. The meaning of the key is that under $dept$, all values for the element sid are distinct. Also, consider the XML inclusion dependency(XID)[11] $\Upsilon_1(dept, (\{project/sid\} \subseteq \{enroll/sid\}))$ on D_1 meaning that the students who have project must be enrolled students. We see that both key and inclusion dependency are satisfied by the document T_1 in Fig.1(b). In case of key, the values of the element sid in $dept$ are $(v_3 : sid : 009), (v_5 : sid : 001)$ and $(v_8 : sid : 007)$ are value distinct. For inclusion dependency, for every values of sid in $project$, there is a value for sid

---

* This research is supported with Australian Research Council(ARC) Discovery(DP) Fund.

D. Ślęzak et al. (Eds.): DTA 2009, CCIS 64, pp. 57–65, 2009.

in *enroll*. Now consider the XML foreign key(XFK) [11] $F_1(dept, (\{project/sid\} \subseteq \{enroll/sid\}))$ on D_1. We see that the foreign key is also satisfied because both key and inclusion dependency are satisfied. Note that the DTD D_1 also conforms to the document T_1.

Now we transform the DTD D_1 and its conforming document T_1 to the DTD D_2 in Fig.2(a) and the document T_2 in Fig.2(b). In transformation, we use a new element *stud* to push down the structure (sid, cid^*). We term this transformation as *expand*. Now we want to know whether the key $k_1(dept, \{enroll/sid\})$ and XID $\Upsilon_1(dept, (\{project/sid\} \subseteq \{enroll/sid\}))$ need any transformation and thus whether they are valid after transformation. We observe that the paths of both key and XID need transformation as a new element *stud* is added in the DTD.

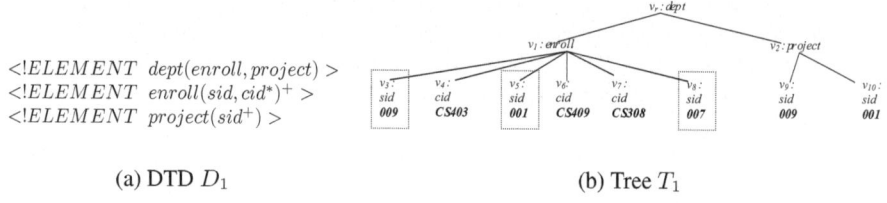

```
<!ELEMENT  dept(enroll, project) >
<!ELEMENT  enroll(sid, cid*)+ >
<!ELEMENT  project(sid+) >
```

(a) DTD D_1 (b) Tree T_1

Fig. 1. Source

We transform k_1 to $k_2(dept, \{enroll/stud/sid\})$ adding new element *stud* in the middle of $field$. In similar way, we transform Υ_1 to $\Upsilon_2(dept, (\{project/sid\} \subseteq \{enroll /stud/sid\}))$. We see that both k_2 and Υ_2 are valid on the DTD D_2 because they follow the path of D_2. As both key and XID are transformed and valid, we transform the XFK $F_1(dept, (\{project/sid\} \subseteq \{enroll/sid\}))$ to $F_2(dept, (\{project/sid\} \subseteq \{enroll/stud/sid\}))$ which is also valid on D_2.

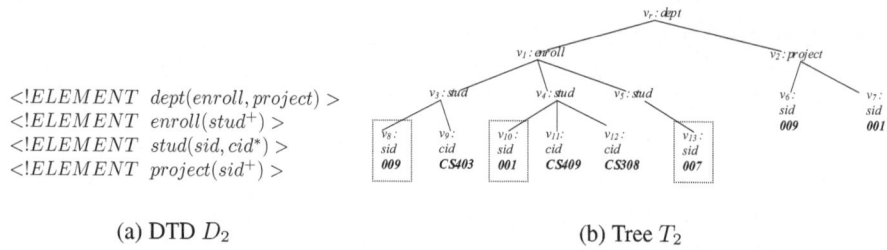

```
<!ELEMENT  dept(enroll, project) >
<!ELEMENT  enroll(stud+) >
<!ELEMENT  stud(sid, cid*) >
<!ELEMENT  project(sid+) >
```

(a) DTD D_2 (b) Tree T_2

Fig. 2. Target

Observation 1. *Transformation of key, XID and XFK needs to be determined when the DTD is transformed.*

Example 2. In the previous example, we showed how to transform key, XID and thus XFK. In this example, we show that though key, XID or XFK are transformed, as key is not preserved so XFK is also not preserved. By key preservation, we mean that if a key is satisfied by the document, after the transformation, the transformed key is also satisfied by the transformed document. Same is with the XID and XFK preservations.

Consider the DTD D_1 in Fig.1(a) and its conforming document T_1 in Fig.1(b). We want to transform D_1 and T_1 to D_3 in Fig.3(a) and T_3 in Fig.3(b). We see that in D_1, the courses are nested with each student. In transformation, we distribute the course id *cid* with student id *sid*. We term this transformation as *unnest*. After the transformation, we see that there is no change of paths in the DTD and thus there is no need to change the key $\Bbbk_1(dept, \{enroll/sid\})$, the XID $\Upsilon_1(dept, (\{project/sid\} \subseteq \{enroll/sid\}))$ or the XFK $F_1(dept, (\{project/sid\} \subseteq \{enroll/sid\}))$. Now we want to see whether the key, the XID or the XFK is preserved.

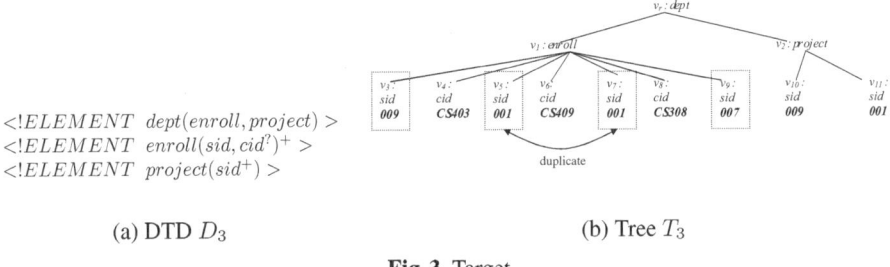

$$<!ELEMENT\ \ dept(enroll, project)>$$
$$<!ELEMENT\ \ enroll(sid, cid^?)^+ >$$
$$<!ELEMENT\ \ project(sid^+) >$$

(a) DTD D_3 (b) Tree T_3

Fig. 3. Target

We observe that the key $\Bbbk_1(dept, \{enroll/sid\})$ (we can say $\Bbbk_3(dept, \{enroll/sid\})$ because it is now on the DTD D_3) is not satisfied by the transformed document T_3 because there there are two values ($v_5 : sid : 001$) and ($v_7 : sid : 001$) for the element *sid* under node $v_1 : enroll$ which are not distinct. However, the XID Υ_1 (we can say Υ_3) is preserved. As the key is not preserved, so the XFK F_1 (we can say F_3) is also not preserved.

Observation 2. *The XFK may not be preserved after the transformation.*

While addressing the problem, we consider the following contributions. *First*, we define the XML foreign key(XFK) and XML Inclusion Dependency (XID) on the XML Document Type Definition (DTD)[16] and their satisfactions using a novel concept called *tuple*. *Second*, we show the transformation on the definition of XFK and XID using basic transformation operations. We also check whether the transformation makes the transformed XFKs and XIDs valid. *Last*, we study whether the transformed XIDs and XFKs are preserved by the transformed XML documents using important operators with sufficient conditions.

2   Transformation on Referential Integrity Constraints

In this section, we study the transformation on XID and XFK. For detailed definitions of XID and XFK, we refer our research in[11,12]. Given a DTD D and a document T such that T conforms to D as $T \in D$, a transformation τ transforms D to $\bar{D}$ and T to $\bar{T}$, denoted by $\tau(D, T) \rightarrow (\bar{D}, \bar{T})$. The problem whether $\bar{T}$ conforms to $\bar{D}$ was investigated in [15]. We investigate how the transformation affects the properties of an

XID Υ defined on D. More formally, given D, T, Υ and a transformation τ such that $\Upsilon \sqsubseteq D \wedge T \Subset D \wedge T \prec \Upsilon$, and $\tau(D, T, \Phi) \rightarrow (\bar{D}, \bar{T}, \bar{\Upsilon}))$, we would like to know what $\bar{\Upsilon}$ is, whether $\bar{\Upsilon}$ is valid on $\bar{D}$, and whether $\bar{T}$ satisfies $\bar{\Upsilon}$.

2.1 Transformation on XID

We now define the transformation of XID. In defining the transformation, we need to refer to the DTD type structure g and the paths w of an XID. For space reason, we refer to the research in[15] for detailed definitions on DTD structure and transformation operations.

Transformation on XID using *Rename* **Operator.** The rename operators changes the element name with a new element name. As the path on the DTD is changed using rename operator, the XID defined on the DTD is also transformed. Formally, let $w = e_1/\cdots/e_k/e_{k+1}/\cdots/e_m$. If the transformation $\tau = rename(e_k, \bar{e}_k)$, then $\tau(w) = \bar{w} = e_1/\cdots/\bar{e}_k/e_{k+1}/\cdots/e_m$.

For example, consider the XID $\Upsilon_1(dept, (\{project/sid\} \subseteq \{enroll/sid\}))$ on the DTD D_1 in Fig.1(a). If we use $rename(sid, studentID)$, then the path P and the path R are transformed and the XID is transformed to $\bar{\Upsilon}_1(dept, (\{project/studentID\} \subseteq \{enroll/studentID\}))$.

Transformation on XID using *Expand* **Operator.** Let $w = e_1/\cdots/e_{k-1}/e_k/e_{k+1}/\cdots/e_m$ be a key path, g be the target of the transformation in $\beta(e_{k-1})$ and $e_k \in g$ (thus $e_k \in \beta(e_{k-1})$ too), and $\tau = expand(g, e_{new})$ be the operator that pushes g one level away from the root. The operator τ transforms $w = e_1/\cdots/e_{k-1}/e_k/e_{k+1}/\cdots/e_m$ to a few cases depending on whether τ adds to the end of Q (i.e., $w = Q$ and $m = k-1$) and whether, in case of appending to the end of Q, all P paths and R paths are pushed down (i.e. $\forall w \in P(beg(w) \in g)$ and $\forall w \in R(beg(w) \in g)$). A path is said **pushed down** means that a new element is added to the beginning of the path.

(1) If w is a dependent P path or w is a referenced R path or w is the selector Q path and $m > (k-1)$ (τ is not adding to the end of Q), then $\tau(w) \rightarrow \bar{w} = e_1/\cdots/e_{k-1}/e_{new}/e_k/e_{k+1}/\cdots/e_m$.

   This case is described in Fig.4 using Case 1. In Case 1(a), $w = Q$ and the new element e_{new} is added between e_{k-1} and e_k in path w. In Case 1(b), $w = P_i$ and e_{new} added between e_{k-1} and e_k in path w. Similarly, In Case 1(c), $w = R_i$ and e_{new} added between e_{k-1} and e_k in path w.

(2) If w is the Q path and $m = (k-1)$ and $\forall w \in P(beg(w) \in g) \wedge \forall w \in R(beg(w) \in g)$, then

   (a) Option 1: $\bar{Q} = Q/e_{new}$ and $\forall w \in P(\bar{w} = w)$ and $\forall w \in R(\bar{w} = w)$.

      This is shown in Case 2(option 1) of Fig. 4 where $w = Q$ and e_{new} is added at the end of Q and all P paths and R paths are not changed.

   (b) Option 2: $\bar{Q} = Q$ and $\forall w \in P(\bar{w} = e_{new}/w)$ and $\forall w \in R(\bar{w} = e_{new}/w)$.

      This case corresponds to Case 2(option 2) of Fig. 4 where $w = Q$ and e_{new} is added at the beginning of all P paths and R paths.

   Both options represent reasonable semantics. The choosing from the two options is determined by the user.

(3) (a) If $last(Q) = e_{k-1} \wedge \exists\, w \in P(beg(w) \notin g)$, then $\bar{Q} = Q$ and $\forall\, w \in P$ (if $beg(w) = e_k$ then $\bar{w} = e_{new}/w$ else $\bar{w} = w$). This case is shown in Case 3(a) of Fig. 4 where e_{new} is added at the beginning of some P paths.

(b) If $last(Q) = e_{k-1} \wedge \exists\, w \in R(beg(w) \notin g)$, then $\bar{Q} = Q$ and $\forall\, w \in R$ (if $beg(w) = e_k$ then $\bar{w} = e_{new}/w$ else $\bar{w} = w$). This case is shown in Case 3(b) of Fig. 4 where e_{new} is added at the beginning of some R paths.

For example, consider the XID $\Upsilon_1(dept, (\{project/sid\} \subseteq \{enroll/sid\}))$ on the DTD D_1 in Fig.1(a). Now if we use $expand([sid \times cid^*], stud)$, then the path on R is only transformed and the transformed XID is $\tilde{\Upsilon}_1(dept, (\{project/sid\} \subseteq \{enroll/stud/sid\}))$ which is valid on the DTD D_2 in Fig.2(a). In this case, the Case 1(c) in Fig.4 is used in transformation.

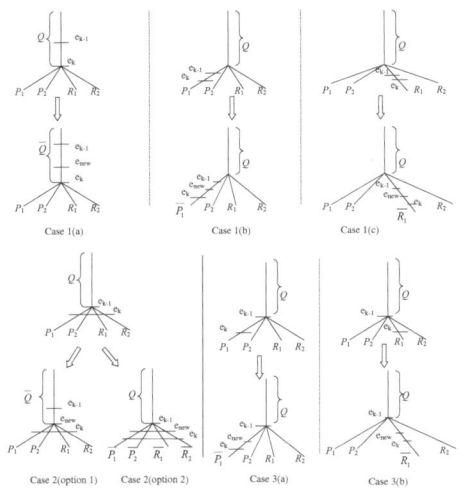

Fig. 4. Transformation on XID using $expand$ operation

Now consider another transformation $expand([enroll \times project], courseDB)$, then we have two options to transform the XID $\Upsilon_1(dept, (\{project/sid\} \subseteq \{enroll/sid\}))$. In first option(using Case 2 and option 1 in Fig.4), we transform the selector by adding the new element $courseDB$ at the end of path $dept$ as $dept/courseDB$ and the transformed XID is $\tilde{\Upsilon}_1(dept/courseDB, (\{project/sid\} \subseteq \{enroll/sid\}))$. In second option(using Case 2 and option 2 in Fig.4), we transform both P path $project/sid$ and R path $enroll/sid$ by adding the new element in the beginning and the transformed XID is $\tilde{\Upsilon}_1(dept, (\{courseDB/project/sid\} \subseteq \{courseDB/enroll/sid\}))$.

Transformation on XID using $Collapse$ **Operator.** If $e_{k+1} \in \beta(e_k)$ and $\tau = collapse(e_k)$, then any path $w = e_1/\cdots/e_{k-1}/e_k/e_{k+1}/\cdots/e_m$ where $(k-1) \geq 1$ is transformed as $\tau(w) \to \bar{w} = e_1/\cdots/e_{k-1}/e_{k+1}/\cdots/e_m$. We note that $collapse(e_k)$ is not applicable if $\beta(e_k) = Str$ and $e_k = \rho$ where ρ is the root.

Theorem 1. *In XID* $\Upsilon(Q, (\{P_1, \cdots, P_n\} \subseteq \{R_1, \cdots, R_n\}))$ *transformation,* Q*, or* P*, or* R *can be transformed, but not* Q, P *or* Q, R *are not transformed together.*

Theorem 2. *Let τ be a transformation defined above such that $\tau(\Upsilon) = \bar{\Upsilon}$. Then $\bar{\Upsilon}$ is valid on $\bar{D}$, denoted as $\bar{\Upsilon} \sqsubset \bar{D}$.*

2.2 Transformation on XFK

As we already mentioned earlier that the XFK $F(Q, (\{P_1, \cdots, P_n\} \subseteq \{R_1, \cdots, R_n\}))$ is defined using the XID $\Upsilon(Q, (\{P_1, \cdots, P_n\} \subseteq \{R_1, \cdots, R_n\}))$ and the XML key $\Bbbk(Q, \{R_1, \cdots, R_n\})$, thus in transforming the XFK, we use the rules of transformation on XID described in the previous subsection and we need to know whether there is a need to transform the key as result of XID transformation. We see that in both XID $\Upsilon(Q, (\{P_1, \cdots, P_n\} \subseteq \{R_1, \cdots, R_n\}))$ and XML key $\Bbbk(Q, \{R_1, \cdots, R_n\})$, the selector path Q and the dependent paths $R_1, \cdots, R_n$ are the same. Thus the rules of transformations on XML key are the same as the rules of transformations on XID except the transformations on the paths of P.

For example, consider the XFK $F_1(dept, (\{project/sid\} \subseteq \{enroll/sid\}))$ on the DTD D_1 in Fig.1(a). If we use the transformation $expand([enroll_\times project]$, $courseDB)$, then we transform the XID $\Upsilon_1(dept, (\{project/sid\} \subseteq \{enroll/sid\}))$. We also need to transform the key $\Bbbk_1(dept, \{enroll/sid\})$. In this case, we have two options like XID transformation. In option 1, we transform as $\bar{\Bbbk}_1(dept/courseDB, \{enroll/sid\})$. In option 2, we transform as $\bar{\Bbbk}_1(dept, \{courseDB/enroll/sid\})$.

Theorem 3. *For the XFK $F(Q, (\{P_1, \cdots, P_n\} \subseteq \{R_1, \cdots, R_n\}))$ where the XID is $\Upsilon(Q, (\{P_1, \cdots, P_n\} \subseteq \{R_1, \cdots, R_n\}))$ and the key is $\Bbbk(Q, \{R_1, \cdots, R_n\})$, Q of both key and XID can be transformed or R of both key and XID can be transformed or P of XID can be transformed.*

Theorem 4. *Let τ be a transformation defined above such that $\tau(F) = \bar{F}$. Then $\bar{F}$ is valid on $\bar{D}$, denoted as $\bar{F} \sqsubset \bar{D}$.*

3 Preservation of XML Referential Integrity Constraints

First, we investigate how a transformation affects the satisfaction of transformed XID. More specifically, given $\tau(D, T, \Upsilon) \rightarrow (\bar{D}, \bar{T}, \bar{\Upsilon}) \wedge \bar{\Upsilon} \sqsubset \bar{D}$, we investigate whether $\bar{T}$ satisfies $\bar{\Upsilon}$ as $\bar{T} \prec \bar{\Upsilon}$.

Definition 1 (XID Preservation). *Given the transformations on D, T, Υ as $\tau(D, T, \Upsilon)$ $\rightarrow (\bar{D}, \bar{T}, \bar{\Upsilon}) \wedge \bar{\Upsilon} \sqsubset \bar{D}$, if $T \prec \Bbbk$ and $\bar{T} \prec \bar{\Upsilon}$, we say that Υ is preserved by the transformation τ.* $\qquad\square$

3.1 Preservation of XID

We now recall the definition of XID $\Upsilon(Q, (\{P_1, \cdots, P_n\} \subseteq \{R_1, \cdots, R_n\}))$ satisfaction which requires that for every P-tuple, there must have a R-tuple under the selector node Q. Thus when we check the preservation of XID, we need to check whether the condition of the XID satisfaction is violated when the document is transformed.

Preservation of XID using *Rename* **operation.** The rename operator only changes the element name with a new element name in DTD. Thus the path involving the element name to be renamed is also transformed and as a result, XID is also transformed.

But the P-tuple or R-tuple is not changed. Also there is no deletion of either P-tuple of R-tuple. So the transformed XID is satisfied by the transformed document.

Preservation of XID using $UnNest$ **and** $Nest$ **operation.** The *unnest* operator transforms the nested structure of a document to the flat structure. In this case,we recall the example where the source tree $T = (\rho(A : 1)(B : 2)(B : 3)(A : 4)(B : 5))$ is transformed to $\bar{T} = (\rho(A : 1)(B : 2)(A : 1)(B : 3)(A : 4)(B : 5))$ using $unnest(B)$ operation. We see that the number of A element is increased in the transformed document. However, there is no loss of information in the transformation. As the XID satisfaction requires that for every P-tuples, there must be a R-tuple, there is no violation XID satisfaction for the transformed document.

In case of *nest* operation, the source tree $T = (\rho(A : 1)(B : 2)(A : 1)(B : 3)(A : 4)(B : 5))$ is transformed to $\bar{T} = (\rho(A : 1)(B : 2)(B : 3)(A : 4)(B : 5))$ using $nest(B)$. We see that the element A with the same values are nested with B elements. Thus there is no violation of XID satisfaction property for the transformed document. We note here that there is no transformation of XID definition using both *unnest* and *nest* operations.

Preservation of XID using $Expand$ **and** $Collapse$ **operation.** The *expand* operator pushes a structure with a new element for semantics. We showed the the transformations on XID using *expand* operation in the previous section. Regardless of different transformation rules and options in transforming XID, there is no change of either P-tuples or R-tuples in the document. The transformed XID is satisfied by the the transformed document and thus XID is preserved.

In case of *collapse*, an element is deleted from the DTD and the document as well. Note that there *collapse* operation is not applicable where the element is the root of the document and $\beta(e) = Str$. After *collapse*, there is no change to P-tuples or R-tuples. The transformed XID is satisfied by the transformed document and thus XID is preserved.

3.2 Preservation of XFK

We recall the definition of XFK $F(Q, (\{P_1, \cdots, P_n\} \subseteq \{R_1, \cdots, R_n\}))$ satisfaction which requires two things: (a) for every P-tuple, there must be a R-tuple under the selector node Q node and (a) all the R-tuples are distinct under the selector node Q. The first requirement means the XID satisfaction and the second requirement means the XML key satisfaction. We already showed how XID is preserved for important transformation operations. Now we show XFK preservation using the same transformation operators.

In XFK preservation, we need both the XID and the XML key preserved. In the previous subsection, we showed the XID preservation. The *rename, nest,* and *collapse* are XID preserving. The XML key is also preserved using *nest, collapse* and *rename* operators because there is no change of R-tuples under the selector nodes for path Q. Although *unnest* and *expand* operators are XID preserving, but they are key preserving with some sufficient conditions. We now study the XFK preservations with *unnest* and *expand* operations.

Preservation of XFK using $UnNest$ **and** $Expand$ **operations.** We already showed that the *unnest* operation is XID preserving. In *unnest* operation, although the XID is preserved, but the XML key is not preserved if some conditions are not satisfied. Thus XFK is also not preserved using *unnest* operation if some conditions of key preservation are not satisfied. We illustrate the condition using the following example.

The *unnest* operator spreads the hedge of the structure g_1 to the hedges of the structure g_2. For example, given $\beta(e) = [A_{\times}B_{\times}[C_{\times}D]^+]^*$ and $T = (e(A : 1)(B : 1)(C : 2)(D : 3)(C : 2)(D : 4)(A : 1)(B : 2)(C : 2)(D : 4))$, the operator $unnest(C_{\times}D)$ spreads the hedge of the structure $A_{\times}B$ to the hedges of the structure $C_{\times}D$ and produces $\beta_1(e) = [A_{\times}B_{\times}[C_{\times}D]]^*$ and $\bar{T} = (e(A : 1)(B : 1)(C : 2)(D : 3)(A : 1)(B : 2)(C : 2)(D : 4)(A : 1)(B : 1)(C : 2)(D : 4))$.

Consider the XML key $\Bbbk(Q, \{R\})$. The $unnest(g_2)$: $[g_1{\times}g_2^+]^+ \to [g_1{\times}g_2]^+$ operator is key preserving if (a) g_1 does not cross the selector Q, and (b) if g_1 crosses R paths, g_2 also crosses some R paths. For example, in the above example, the key is $\Bbbk(e, \{A, B\})$, then it is satisfied by the document T because there are R-tuples $((A : 1)(B : 1))$ and $((A : 1)(B : 2))$ which are distinct. But the after the *unnest* operation, the key is not satisfied by the document $\bar{T}$ because there are two R-tuples as $((A : 1)(B : 1))$ which are vale equivalent in the document $\bar{T}$. Thus the key is not preserved after *unnest* operation.

We showed that the *expand* operator is XID preserving. But *expand* operations is key preserving if some sufficient conditions are satisfied. Thus the *expand* operator is XFK preserving if some sufficient conditions are satisfied.

4 Conclusions

We studied the preservations of both XML inclusion dependency and XML foreign key. We found that although inclusion dependency is preserved for transformation operations, but foreign key is not preserved for some important transformation operations. We then identified sufficient conditions for preservations. Our study on preserving referential integrity for XML is towards the data integration in XML with integrity constraints.

References

1. Fankhouser, P., Klement, T.: XML for Datawarehousing Chances and Challenges. In: Kambayashi, Y., Mohania, M., Wöß, W. (eds.) DaWaK 2003. LNCS, vol. 2737, pp. 1–3. Springer, Heidelberg (2003)
2. Zamboulis, L., Poulovassilis, A.: Using Automed for XML Data Transformation and Integration. In: DIWeb, pp. 58–69 (2004)
3. Zamboulis, L.: XML data integration by graph restructuring. In: Williams, H., MacKinnon, L.M. (eds.) BNCOD 2004. LNCS, vol. 3112, pp. 57–71. Springer, Heidelberg (2004)
4. Poggi, A., Abiteboul, S.: XML Data Integration with Identification. In: Bierman, G., Koch, C. (eds.) DBPL 2005. LNCS, vol. 3774, pp. 106–121. Springer, Heidelberg (2005)
5. Buneman, P., Fan, W., Simeon, J., Weinstein, S.: Constraints for Semistructured Data and XML. In: SIGMOD Record, pp. 47–54 (2001)
6. Fan, W.: XML Constraints: Specification, Analysis, and Applications. In: DEXA, pp. 805–809 (2005)
7. Fan, W., Simeon, J.: Integrity constraints for XML. In: PODS, pp. 23–34 (2000)

8. Fan, W., Libkin, L.: On XML Integrity Constraints in the Presence of DTDs. Journal of the ACM 49, 368–406 (2002)
9. Vincent, M., Schrefl, M., Liu, J., Liu, C., Dogen, S.: Generalized inclusion dependencies in XML. In: Yu, J.X., Lin, X., Lu, H., Zhang, Y. (eds.) APWeb 2004. LNCS, vol. 3007, pp. 224–233. Springer, Heidelberg (2004)
10. Karlinger, M., Vincent, M., Scherefl, M.: On the Definition and Axiomitization of Inclsuion Dependency for XML. In: Tecnical Report, No.07/02,Johanne Kepler University (2007)
11. Shahriar, M.S., Liu, J.: On Defining Referential Integrity for XML. In: IEEE International Symposium of Computer Science and Its Applications(CSA), pp. 286–291 (2008)
12. Shahriar, M.S., Liu, J.: On Defining Keys for XML. IEEE CIT 2008, Database and Data Mining Workshop, DD 2008, 86–91 (2008)
13. Shahriar, M. S., Liu, J.: Preserving functional dependency in XML data transformation. In: Atzeni, P., Caplinskas, A., Jaakkola, H. (eds.) ADBIS 2008. LNCS, vol. 5207, pp. 262–278. Springer, Heidelberg (2008)
14. Shahriar, M. S., Liu, J.: Towards the Preservation of Keys in XML Data Transformation for Integration. In: COMAD, pp. 116–126 (2008)
15. Liu, J., Park, H., Vincent, M., Liu, C.: A Formalism of XML Restructuring Operations. In: Mizoguchi, R., Shi, Z.-Z., Giunchiglia, F. (eds.) ASWC 2006. LNCS, vol. 4185, pp. 126–132. Springer, Heidelberg (2006)
16. Tim Bray, Jean Paoli, and C. M. Sperberg-McQueen, Extensible Markup Language (XML) 1.0., World Wide Web Consortium, W3C (Febuary 1998), http://www.w3.org/TR/REC-xml
17. Thompson, H.S., Beech, D., Maloney, M., Mendelsohn, N.: XML Schema Part 1:Structures, W3C Working Draft (April 2000), http://www.w3.org/TR/xmlschema-1/

Know-Ont: Engineering a Knowledge Ontology for an Enterprise*

Harshit Kumar and Pil Seong Park

Computer Science Department, University of Suwon, South Korea
{harshitkumar,pspark}@suwon.ac.kr

Abstract. Research in the field of knowledge representation system is usually focused on methods for providing high-level descriptions of the world that can be effectively used to build intelligent applications. This paper shares the development process of ontology which involves preparing questionnaires, design decisions, and interviewing key persons. This is an ongoing work resulting in the development of ontology for knowledge management in an enterprise. The final result is a knowledge ontology, coined as Know-Ont, is a collection of concepts and their related properties from maintenance and new product design domain that fit together to process and store knowledge thus making it available for later reuse.

Keywords: Knowledge Management, Ontology, Semantic Web, AI.

1 Introduction

Knowledge representation [11] developed as a branch of artificial intelligence is defined as the science of designing computer systems to perform tasks that would normally require human intelligence. Knock et al. [6] states that *the single most important factor that ultimately defines the competitiveness of an organization is its ability to acquire, evaluate, store, use and discard knowledge and information.* Bradley et al. [1] estimate that, knowledge is currently doubling every 18 months; therefore knowledge should be managed efficiently and effectively so that it can be put to re-use.

Our hypothesis is based on the following: if a system can somehow understand the user requirements and be aware of the situation a user is working on, it might probably be able to put the knowledge management system to use: adapt itself to meet different user's requirements, automatically monitor the knowledge process and extract reusable knowledge, proactively provide right knowledge at the right time, and so on. But how to achieve such a system still remains at large. Using ontologies is one of the ways; the first step in this direction is to develop knowledge ontology for industrial domain. The principal contribution of this paper is to present and discuss the various steps followed during the process of ontology development; we studied

---

* This work was supported by the GRRC program of Gyeonggi province. [GRRC Suwon2009-A2, Research on knowledge-based context-aware information system].

D. Ślęzak et al. (Eds.): DTA 2009, CCIS 64, pp. 66–73, 2009.

business cases provided by our industrial partners (name withheld due to privacy issues), prepared questionnaires, conducted interviews, which resulted in the definition of concepts and its related properties. Knowledge ontology, Know-Ont is envisioned as a final result. To the best of our knowledge, this work is one-of-its kind, and no ontology exists for modeling knowledge management in industrial scenarios. Though there are few works that appear similar to ours [8], however our approach and methodology serves a different purpose. A working prototype of the system is implemented and queried using SPARQL[1].

Section 2 provides introduction to ontology in industrial domain and also explain how our work is different from other existing ontologies. Section 3 presents the business cases and requirement analysis, which is followed by the discussion about the resulting knowledge ontology Know-Ont in Section 4. The final section compromises future work and conclusion.

2   Ontologies in Industrial Domain

Ontology is an explicit specification of conceptualization [4] i.e. ontology is a description of concepts and relationships that can exist between them. The concepts can be determined manually or through an experimental process. In this paper, we use software engineering principle to determine concepts and their associated relationship. Some of the concepts are borrowed from existing ontologies like GEO[2], FOAF[3] etc. It is normal to borrow concepts from other ontologies, this puts ontologies to re-use.

There are numerous ontologies available on the web but very few of them provide explicit details about the process followed that lead to its development. These ontologies model process flow in industrial scenarios, for example there is a manufacturing system engineering ontology [9], ontology for virtual enterprise integration [3], ontology for supply chain [10], and so on. We on the other hand advocate use of ontologies for knowledge management in a different context. Our objective is to provide reusability of stored information in an efficient and effective way, so that right information is available to the right person at the right time.

Mason [8] is a popular ontology for industrial scenarios that models manufacturing domain. One of the applications of this ontology is automatic cost estimation; If someone tries to envision an application for Know-Ont, it would be search for persons who can do a particular job, time duration required to finish a job, searching relevant documents similar to a job in hand, and search for similar activities etc.

3   Business Cases and Requirement Analysis

This section details the steps to elicit information from 3 industrial partners, each from different domain viz. electrical engineering industry, aerospace industry, and

---

[1] http://www.w3.org/TR/rdf-sparql-query/ - SPARQL is a query language for querying RDF repositories.

[2] http://www.w3.org/2003/01/geo/ - an ontology for spatial modeling like country, cities etc.

[3] http://xmlns.com/foaf/spec/ - an ontology for describing a person.

engineering consultancy. Each industrial partner was asked to identify target processes in their respective domains where knowledge enhancement is needed. A questionnaire was prepared and distributed among the industrial partners for their feedback.

Industrial partners were asked to provide a business case based on the following format: description, description of process, definition of quantifiable business objectives, expectation of improvement after the implementation, use case modeling of the processes. For the elaboration of both the business case description and requirements description, personal interviews were conducted.

3.1 Analysis of Business Cases

In this section, we first present results from the analysis of business cases and questionnaire, i.e. identification of areas or processes to improve, definition of quantitative objectives to achieve. Further, we will lay down the basis for the proposed ontology.

Industrial Partner A identified two main processes: product design and maintenance service. Both processes involve manual search of documents which is a tedious process, for ex: product design activity (**Activity**[4]) generally involves improvising an existing product (**Artefact**) by introducing current market requirements. For improvising an existing product, its previous design methodology, sketches, drafts, (**Document**) etc. are needed to be readily available; moreover, searching for appropriate and available staff (**Person**) who can carry out the design job in stipulated time (**Time**) is also need to be determined. Similar type of issues exists for the maintenance department also. The design process involves most of the departments (**Department**) of the company (Technical office, Quality Department, Marketing, etc) as well as external organizations (commercial organizations, labs, etc).

Industrial partner B identified maintenance process. This process can be of two types: corrective maintenance i.e. the process launches after a customer request arrives. Another type of maintenance is scheduled maintenance like AM (Annual Maintenance). They envisioned the following objectives from the proposed solution: improved network for co-operative networking, personalized solutions, and remote maintenance (**Location**) by providing information readily (semantic search) everywhere and anywhere.

The third industrial Partner C selected maintenance process and project development as two main processes. The actors involved in the maintenance process are distributed in several subsidiaries or locations (**Location**) across the country. Finding the right available person for the right task can be a daunting task taking into account different skills (**Role**), availability, and location. Developing a new project generally involves improvising an already existing project; similar requirements were posted by industrial partner A also.

It is apparent from the above discussion; one common process that emerges out of business cases is Maintenance. On careful examination of processes "Product Design" and "Project Development", we found out that they reflect similar objective i.e. new

---

[4] Text in bold reflects the concepts that were extracted from the business cases – Here we are only showing a partial list of concepts as an illustration.

product design. We therefore decided to provide proof of concept for Maintenance and Product Design business cases. Note that, the proposed ontology is a generic one which can be extended to other business cases also. Such ontology in the literature is referred as upper ontology[5]. Based on the evaluation of business cases, we define a set of concepts (refer Table 1), which were either borrowed from other ontologies developed by us or from ontologies developed by others like FOAF, Time, Geo etc.

Table 1. A partial list of concepts with their definitions for consideration to industrial partners. A complete list is not provided because of number of page issues, however, important concepts generated from the analysis of business cases are listed below.

Concept Name	Source	Definition
Location	Geo	An abstract concept to describe any locatable entity
Resource	Self	A generic concept for any means that is used in an activity.
Document Resource	Self	Anything that can be regarded a document such as text files, spreadsheets, rtf documents,
Service	Self	Denotes anything described by a well-known, published interface.
Person	FOAF	A generic concept that describes people
Department	Self	Represents affiliation of a person
Profile	Self	Profession description of a person
Activity	Self	Describes everything a person/process has done, is doing, or will be doing (assigned tasks) in order to fulfill a goal.
Role	Self	Defines a role performed by the engineer
Task	Self	An activity is composed of several tasks
Artefact	Self	Final product
Time	Time[6]	To track the start, end and intermediate stages of a process
Process	Self	A process is composed of several activities

4   Knowledge Ontology – Know-Ont

In this section, we lay the foundation of Know-Ont ontology; explain the various concepts and their associated properties. A very simple prototype of Know-Ont is developed using Protégé[7]. Protégé is an ontology editor that provides interface for inputting instances and querying them using SPARQL. Our basic objective is to model all the available information as instances in ontology and making it available

---

[5] http://en.wikipedia.org/wiki/Upper_ontology_(information_science) – definition of Upper Ontology.

[6] http://www.w3.org/TR/owl-time/ - an ontology for temporal concepts.

[7] Protégé - http://protege.stanford.edu/ - is an ontology editor.

for later re-use. Section 4.1 explains the specifications of Know-Ont followed by Section 4.3 that demonstrates how Know-Ont can be put to use for searching information based on an example use case in Section 4.2.

4.1 Know-Ont Specifications

Based on the feedback from industrial partners and meticulously going through the business cases, we arrived at the Know-Ont ontology shown in Fig. 1.

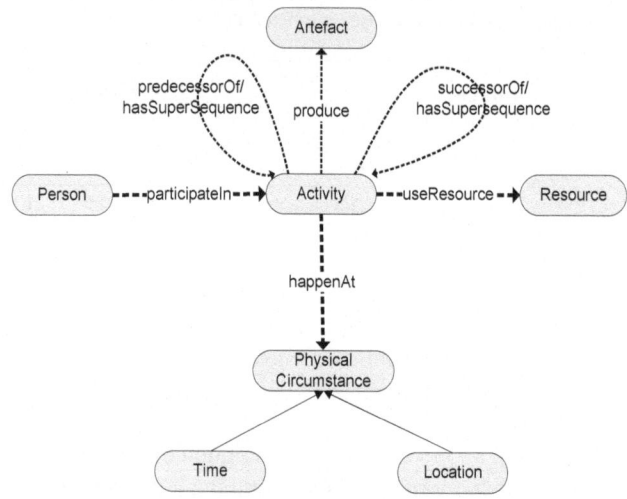

Fig. 1. Overview of Knowledge Ontology: Know-Ont

The central concept is an *activity*, which represents a knowledge based activity and is very general to include any sort of processes in an industrial domain. Each activity is performed by a group of human beings, which is modeled as a *person* concept and can be imported from the FOAF ontology. Other concepts like department, skill, etc are also imported from FOAF. An *Activity* is related to itself as a predecessor/successor activity to represent hierarchy, as well as a hasSubsequence/hasSupersequence activity to represent time sequence. The concept *Activity* can be made domain specific by extending into more specific concept which may vary from case to case. For example, we extend concept *Activity* into several specific sub classes viz. Product Design Activity, Maintenance Activity, etc. as shown in Fig. 2.

Every *Activity* has some physical circumstances associated with it, like, *Location* and *Time*. *Physical Circumstance* is a Concept. An *Activity* is related to *Physical Circumstance* through the object property happenAt. Result of an *activity* is ultimately an *Artefact* (or Product). Know-Ont has a layered architecture, so *Artefact* represents a part of core ontology and there is a further classification of *Artefact* which depends on the application domain. The object property particpateIn defines the relationship between the *Person* and *Activity*.

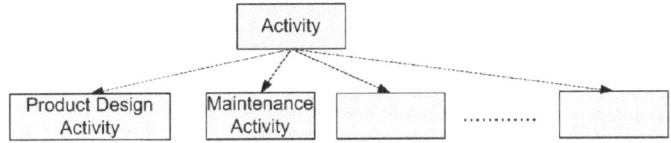

Fig. 2. Extension of Activity concepts into domain specific concepts

Each *Activity* uses some *Resource* (such as Document, Manuals, URLs, tools, etc) to produce an *Artefact*. Therefore, a concept *Resource* is defined, which is related to the *Activity* through the useResource object property. To make the ontology domain specific, the concept Resource can be further extended into *Document, Tool*, etc. as shown in Fig. 3.

Note the difference between concepts *Resource* and *Artefact*; Resource is something that is used by an activity for the production of a product (referred as *Artefact*). This means that in one industrial environment, scissor can be a resource and in other it can be an Artefact. We plan to re-use existing product ontology. There are numerous product ontologies [5, 7] available; work to identify which product ontology suits out requirement is in progress.

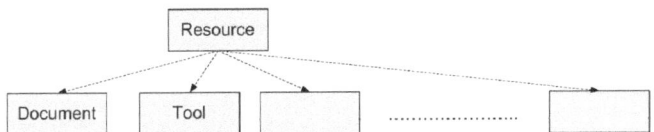

Fig. 3. Domain specific ontology extension to core concept resource

4.2 A Use-Case Representation of Know-Ont Ontology

Fig. 4. presents a use case using Know-ont ontology, which shows two activities, persons involved in the activity, artefact produced and tools used to produce the artefact. It also shows relationship between two activities. For simplicity, we have used OOPS model to represent class name and instance name. For ex: concept Activity is represented as a Class and the term "Know-Ont Concept development" is represented as an instance of Activity concept. It can be clearly seen that "Know-Ont Context Extraction" activity is successor of "Know-Ont Concept development" activity. Note the difference between precedence and subsequence object property; if an activity A is 'predecesorOf' activity 'B', it means, activity 'A' executes and terminates following which activity B starts. Whereas, if an activity A 'hasSubsequence' activity B, it means, activity A needs activity B to finish and use generated results to finish itself. Similar difference exists between 'successorOf' and 'hasSupperSequence' object property. Also note that, 'predecessorOf' is owl:inverseOf 'successorOf' object property.

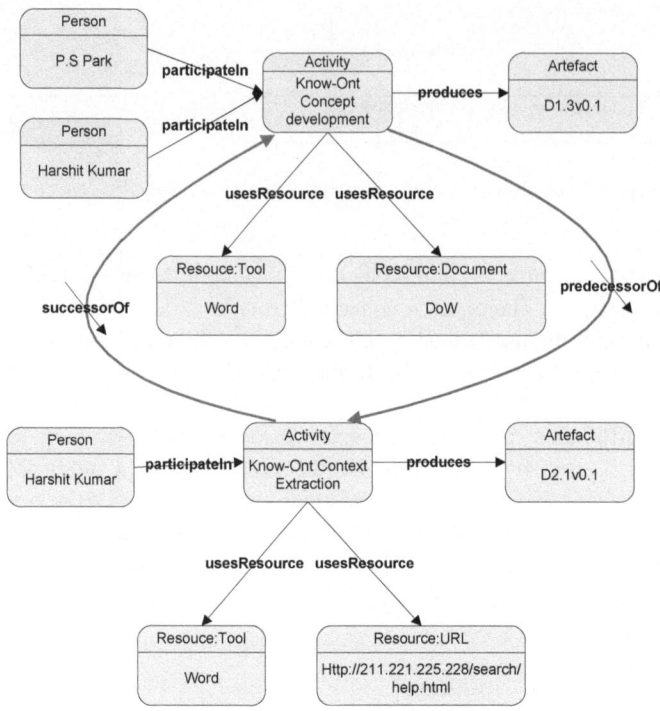

Fig. 4. A use case of Know-Ont ontology represented using OOPS model

4.3   Querying Know-Ont Using SPARQL

The Know-Ont ontology is available to address the following questions: *Is there any similar product exists?* If answer to above question is yes, then, *List the similar products and the degree of similarity, List the classification of existing product, List the actors involved, List the sketches, List the technical specs,*

For ex: Find all the artefacts produced by user "Harshit Kumar". The query is as follows

> SELECT ?person ?activity ?artefact WHERE { ?person rdf:typeOf
> foaf:Person. ?person foaf:name "Harshit Kumar". ?person ko:participateIn
> ?activity. ?activity ko:produces ?artefact. }

The above query will search for tuples that have foaf:name "Harshit Kumar", who has participated in some activity that results in some artefacts, output is shown in Table 2.

Table 2. SPARQL query output, find all the artefacts produced by user "Harshit Kumar"

?person	?activity	?artefact
Harshit Kumar	Know-Ont Concept Development	D1.3v0.1
Harshit Kumar	Know-Ont Context Extraction	D2.1v0.1

5 Conclusion and Future Work

The work presented in this paper deals with engineering a knowledge ontology which can be further re-used for knowledge provisioning. The process to engineer knowledge ontology involves interviewing, meeting with the key persons who execute or deal with the target business process. Knowledge ontology Know-Ont is based on two business cases selected by industrial partners, which are derived from different domains viz. maintenance and product design domain. A very simple prototype has been realized using Protégé and a use case is presented to show the results from the system on which a SPARQL query executes for finding all the artefacts produced by a particular user.

Certainly many works still need to be done, for instance, an interface for users to input data and query. Query results will probably return more than one match; all the matches need to be ranked. For ranking, we must define a similarity measure that can return tangible numbers so that we can differentiate which match is closer to the current context and which one is farther.

References

1. Bradley, K.: Intellectual Capital and the New Wealth of Nations. Business Strategy Review 8(1), 53–62 (1997)
2. Broekstra, J., Kampman, A., van Harmelen, F.: Sesame: An architecture for storing and querying RDF and RDF schema. In: Horrocks, I., Hendler, J. (eds.) ISWC 2002. LNCS, vol. 2342, pp. 54–68. Springer, Heidelberg (2002)
3. Chen, G., Zhang, J.B., Low, C.P., Yang, Z., Ren, W., Zhuang, L.: Collaborative Virtual Enterprise Integration via Semantic Web Service Composition, pp. 1409–1414 (2007)
4. Gruber, T.: Ontology. In: Liu, L., Tamer Özsu, M. (eds.) Entry in the Encyclopedia of Database Systems. Springer, Heidelberg (2008)
5. Hepp, M.: GoodRelations: An Ontology for Describing Products and Services Offers on the Web. In: Gangemi, A., Euzenat, J. (eds.) EKAW 2008. LNCS (LNAI), vol. 5268, pp. 332–347. Springer, Heidelberg (2008)
6. Knock, N., McQueen, R., Corner, J.: The nature of data, information and knowledge exchanges in business processes: Implications for process improvement. The Learning Organization 4(2), 70–80 (1997)
7. Lee, T., Lee, I.H., Lee, S., Lee, S.G., Kim, D., Chun, J., et al.: Building an operational product ontology system. Electronic Commerce Research and Application 5(1), 16–28 (2006)
8. Lemaignan, S., Siadat, A., Dantan, J.-Y., Semenenko, A.: MASON: A proposal for an ontology of manufacturing domain. In: IEEE Workshop on Distributed Intelligent Systems (DIS), pp. 195–200 (2006)
9. Lin, H.K., Harding, J.A.: A manufacturing system engineering ontology model on the semantic web for inter-enterprise collaboration. Computers in Industry 58, 428–437 (2007)
10. Preist, C., Cuadrado, J.E., Battle, S., Williams, S., Grimm, S.: Automated Business-to-Business Integration of a Logistics Supply Chain using Semantic Web Services Technology. In: Gil, Y., Motta, E., Benjamins, V.R., Musen, M.A. (eds.) ISWC 2005. LNCS, vol. 3729, pp. 209–223. Springer, Heidelberg (2005)
11. Sowa, J.F.: Knowledge Representation: Logical, Philosophical, and Computational Foundations. Brooks Cole Publishing, Pacific Grove (1999)

Transformation of Data with Constraints for Integration: An Information System Approach[*]

Md. Sumon Shahriar and Jixue Liu

Data and Web Engineering Lab
School of Computer and Information Science
University of South Australia, SA-5095, Australia
shamy022@students.unisa.edu.au, jixue.liu@unisa.edu.au

Abstract. Transformation of data from different information systems is important and challenging for integration purposes as data can be stored and represented in different data models in different information systems. Moreover, when modeling data in the information systems, integrity constraints on the schemas are necessary for semantics and to maintain consistency purposes. Thus when schemas with conforming data are transformed from heterogeneous information systems, there is a need to transform and preserve semantics of data using constraints. In this paper, we propose how data from different source information systems can be transformed to a global information system. We also review how constraints in data transformation are used in data integration for the purpose of integrating information systems. Our research is towards the handling of semantics using integrity constraints in data integration from heterogeneous information systems.

1 Introduction

As different information systems(IS) use different data models for storing and representing data, transformation of data for exchange and integration purposes [1] is necessary and is a challenging task. Each information system is designed with its own data model and schemas are designed with constraints to convey semantics of data and for maintaining consistency. Historically most popular data model is relational data model. However, much of data is currently being represented and exchanged in XML [7] over the world wide web. Thus transformation of XML data to relational data and transformation of relational data to XML data become necessary for data intensive activities such as data integration, data warehousing, data exchange and data publishing [21,28]. We show the data transformation and integration for heterogeneous information systems in Fig. 1.

The Fig. 1 shows the data transformation and integration architecture for different information systems with different data models. Both local information

---

[*] This research is supported with Australian Research Council(ARC) Discovery Project Fund.

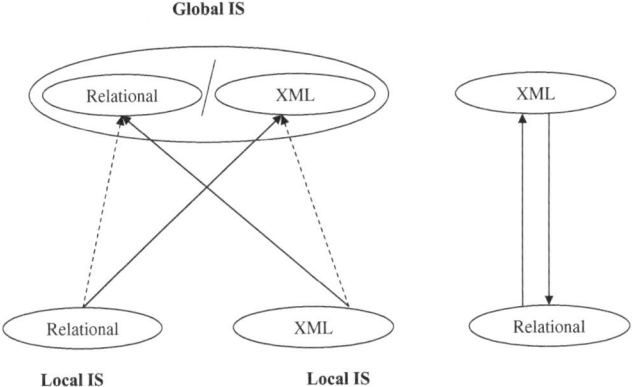

Fig. 1. Data integration from heterogeneous information systems

systems and the global information system can be either in XML or in relational database. When both local information systems and the global information system have different database systems, there is a need to transform schema and data with constraints. We also identify that even for the same database systems(e.g.XML) for both local and global information systems, there is a need to transform schema and data with constraints for preserving semantics.

We study the followings in this paper.

- We review the researches of data transformation and integration with constraints for different data models.
- We then identify the data transformation framework for different data models.
- A data integration framework based on different data models with constraints from different information systems is also proposed.

2 Data Transformation and Integration with Constraints: Overview

We now review the achievements of transformations of data with constraints for different data models in information systems.

2.1 Relational to XML Data Transformation $[R - X]$

With the emergence of XML as data representation and storage format over the web, data transformation and integration from relations to XML became necessary for some purposes such as data exchange and data publising. In relational to XML data transformation, the issue of constraints preservation is studied in [27] where the constraints like primary keys, foreign keys, unique constraints, and not null are considered when a relation is transformed to XML Schema. In publishing data in XML from XML and relational data [28], constraints are also exploited for better query formulation. In data translation [21], both XML and relational schemas are considered with some constraints like nested referential constraints.

2.2 XML to Relational Data Transformation $[X - R]$

Like relational to XML, data transformation and integration from XML to relations also got much attention in past. In [23], constraints(e.g., cardinality constraints, domain constraints, inclusion dependencies etc.) are considered for preservation when XML DTD is transformed to relational schema. In [22], XML keys are transformed to functional dependencies(FDs) for relations and this process is termed as constraint propagation. In [25], how relational keys can be captured from XML keys is shown. Some work ([24,26]) where constraints like cardinality constraints, inclusion dependencies, constraints on DTD etc. are considered where preservation is the issue for XML to relational data transformation perspective.

Discussions: Transformation of XML constraints are very different from relational constraints in the sense that constraints in XML follow the tree structure of data and relational constraints follow flat structure of data. Thus when transforming an XML schema with data to relational schema with data, there is a need to investigate how constraints in XML are transformed to relational constraints with preservation. Similarly, when relational schema with data is transformed to XML schema with data, there is a need to investigate how to transform constraints in relations to constraints in XML and how to preserve constraints by the transformed XML data.

We now discuss the data transformation for homogeneous data model.

2.3 Relational to Relational Data Transformation $[R - R]$

The research works in [10,35,36,37,38] are the examples of data integration in relational database and use the transformation or restructuring of source schemas for schema integration.

Integrity constraints integration for schema integration was studied for heterogeneous databases in [10] and the research was mainly in relational data model.

McBrien and Poulovassilis [35] developed a data integration system in relational model. They use a set of primitive transformations on source schemas to integrate them into global schema. A general framework for transformation of schemas is shown in [37] by them. In [38], they described a formal framework for schema integration that uses a common data model in Entity-Relationship (ER) model. They also proposed a set of transformation operators for schema integration in [38]. They extended their research works and presented an approach named schema evolution [36] in schema transformation and integration.

In pure relational data integration [3], how queries should be affected in the presence of keys and foreign keys on the global schema and no constraints on the source schema is shown. In [15], a data integration system was shown where both source schemas and the global schema can have constraints and then he showed how the source derived global constraints and the original constraints on the global constraints can further be used for answering the query. In [16], how the consistent local constraints on the local schemas are transformed and simplified to the global schema is shown. In relational schema integration, the correspondence between local integrity constraints and the global extensional assertions is investigated in [17]. In [18], the query preserving transformation in data integration system is shown with or without constraints on the global schema. In [19], how the integrity constraints over the global schema can affect the query answering is discussed. In [20], the inconsistency of a data integration system is illustrated when source constraints over the source schemas are not the same as global constraints over the global schema.

2.4 XML to XML Data Transformation $[X - X]$

In recent years, with the massive use of XML over world wide web, the task of data transformation and integration in pure XML is worth to mention. In [2], XML keys and foreign keys are taken into consideration on the XML global schema where source schemas are also in XML. In [30], data from relational sources are integrated to the XML target schema where keys and foreign keys in relations are captured as XML keys and XML inclusion constraints on the target schema using constraint compilation. In XML to XML data transformations and integration ([32,14,9,11,12,13]), how the important XML constraints(e.g. XML keys, XML Functional Dependencies) on the source schema should be transformed and preserved to the target schema is investigated in [5,6].

3   Data Transformation Framework in Heterogeneous Data Model

In data transformation for integration, a source schema with its conforming data is transformed to target schema. A source schema can be defined with integrity constraints to convey semantics of data. When a source schema is transformed to target schema with their conforming data, constraints need to be transformed and preserved. We illustrate these problems using the Fig.2.

We denote a database of an information system as a triple $\Delta = (S, T/I, C)$ where S is schema, T/I is document or instance and C is constraint. We use $\Delta_X = (S_X, T_X, C_X)$ to mean a database in XML and $\Delta_R = (S_R, I_R, C_R)$ to mean a database in relational model.

We denote the transformation as τ that has three sub operations: the schema transformation τ_S, the document or instance transformation $\tau_{T/I}$ and the constraint τ_C.

We now study the effects of transformations on schema, document or instance and constraint.

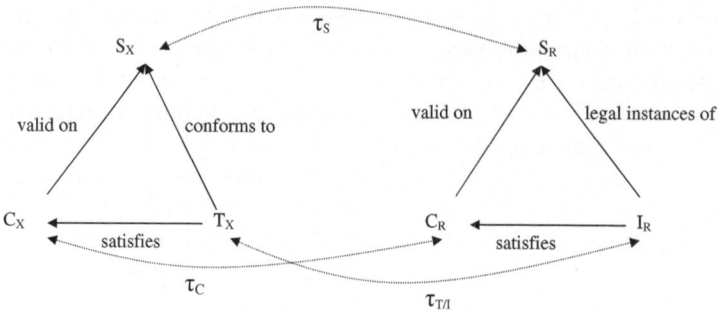

Fig. 2. Data transformation of heterogeneous information systems

3.1   Schema Transformation $[\tau_S(S)]$

In schema transformation, different transformation operations are used. For $\tau(\Delta_R) \rightarrow \Delta_X$, relational tables need to be transformed to different schema definitions in XML such as XML Document Type Definitions(DTD)[7] and XML Schema [8]. Similarly, for $\tau(\Delta_X) \rightarrow \Delta_R$, different schemas in XML are to be transformed to relations.

3.2   Data Transformation $[\tau_{T/I}(T/I)]$

When schemas are transformed, underlying data or documents conforming to the schemas need to be also transformed. For $\tau(\Delta_R) \rightarrow \Delta_X$, flat structure instances of relational tables are to be transformed to tree structure XML documents. Opposite task is needed with the case of $\tau(\Delta_X) \rightarrow \Delta_R$. One property known as *information preservation*[31] is necessary when data is transformed.

3.3   Constraints Transformation $[\tau_C(C)]$

When schemas with data are transformed, the constraints specified on the schemas need to be transformed. For $\tau(\Delta_R) \rightarrow \Delta_X$, constraints(Primary key,unique constraints,functional dependency, foreign key etc.) in relational data are to be transformed to XML constraints(XML absolute key, relative key, XML functional dependency, XML inclusion dependency etc.). When transforming relational constraints to XML constraints, the graph-structured XML schema and the tree-structured XML documents need to be considered. There is also a need to preserve the constraints[23,24,27] in transformations. Similarly, the opposite transformations and preservations are needed when the transformation is $\tau(\Delta_X) \rightarrow \Delta_R$.

4   Data Integration Framework with Constraints in Different Data Models for Information Systems

In data integration, from sources, schema and its conforming data with consistent constraints need to be transformed and integrated to the global site. Moreover, the global information system needs to have its own constraints for data consistency and integrity. We show this framework in the Fig. 3. The global information system is denoted as I^G and the source information systems are denoted as I^{S1} and I^{S2}.

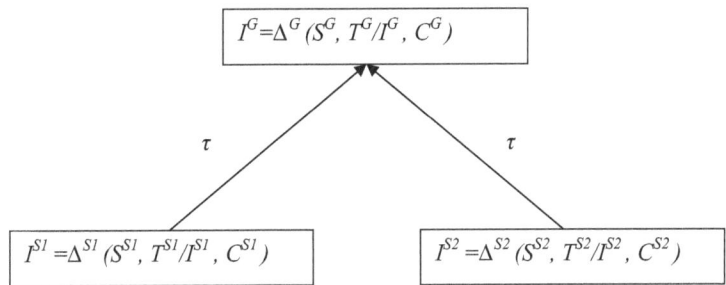

Fig. 3. Data transformation and integration of heterogeneous information systems

In transforming schema S, document or instance T/I and constraint C, there is a need of a set of transformation operations. These operations need to be sufficient to transform a source schema with its conforming data and the constraints on the schema to the global schema.

We identify the following tasks need to be performed in data integration with constraints.

(a) The constraints on the source are to be transformed and preserved to the global site for semantics.
(b) The global constraints need to be consistent with the integrated data from sources.
(c) There is a need to find the non-equivalent constraints between the source derived constraints at the global site and the constraints defined on the global site.
(d) The correspondences between relational constraints and XML constraints need to be investigated as there are many proposals on XML constraints[39] while relational constraints are well established.

5   Conclusions

We reviewed the constraints in data transformation and integration for different data models for information systems. We then showed what should be the architecture for data transformation according to the review. We also showed the data integration framework that used the transformations of schema, data and constraints.

References

1. Lenzerini, M.: Data Integration: A Theoretical Perspective. In: ACM PODS, pp. 233–246 (2002)
2. Poggi, A., Abiteboul, S.: XML Data Integration with Identification. In: Bierman, G., Koch, C. (eds.) DBPL 2005. LNCS, vol. 3774, pp. 106–121. Springer, Heidelberg (2005)
3. Cali, A., Calvanese, D., Giacomo, G.D., Lenzerini, M.: Data Integration under Integrity Constraints. In: Pidduck, A.B., Mylopoulos, J., Woo, C.C., Ozsu, M.T. (eds.) CAiSE 2002. LNCS, vol. 2348, pp. 262–279. Springer, Heidelberg (2002)
4. Amer-Yahia, S., Du, F., Freire, J.: A comprehensive solution to the XML-to-relational mapping problem. In: WIDM, pp. 31–38 (2004)
5. Shahriar, M. S., Liu, J.: Preserving Functional Dependency in XML Data Transformation. In: Atzeni, P., Caplinskas, A., Jaakkola, H. (eds.) ADBIS 2008. LNCS, vol. 5207, pp. 262–278. Springer, Heidelberg (2008)
6. Shahriar, M.S., Liu, J.: Towards the Preservation of Keys in XML Data Transformation for Integration. In: COMAD, pp. 116–126 (2008)
7. Bray, T., Paoli, J., Sperberg-McQueen, C.M.: Extensible Markup Language (XML) 1.0., World Wide Web Consortium, W3C (Febuary 1998), http://www.w3.org/TR/REC-xml
8. Thompson, H.S., Beech, D., Maloney, M., Mendelsohn, N.: XML Schema Part 1:Structures, W3C Working Draft (April 2000), http://www.w3.org/TR/xmlschema-1/
9. Liu, J., Park, H., Vincent, M., Liu, C.: A Formalism of XML Restructuring Operations. In: Mizoguchi, R., Shi, Z.-Z., Giunchiglia, F. (eds.) ASWC 2006. LNCS, vol. 4185, pp. 126–132. Springer, Heidelberg (2006)
10. Ramesh, V., Ram, S.: Integrity Constraint Integration in Heterogeneous Databases: An Enhanaced Methodology for Schema Integration. Information Systems 22(8), 423–446 (1999)
11. Zamboulis, L., Poulovassilis, A.: Using Automed for XML Data Transformation and Integration. In: DIWeb, pp. 58–69 (2004)
12. Zamboulis, L.: XML Data Integration by Graph Restructuring. In: Williams, H., MacKinnon, L.M. (eds.) BNCOD 2004. LNCS, vol. 3112, pp. 57–71. Springer, Heidelberg (2004)
13. Su, H., Kuno, H., Rudensteiner, E.A.: Automating the Transformation of XML Documents. In: WIDM, pp. 68–75 (2001)
14. Erwig, M.: Toward the Automatic Derivation of XML Transformations. In: Jeusfeld, M.A., Pastor, Ó. (eds.) ER Workshops 2003. LNCS, vol. 2814, pp. 342–354. Springer, Heidelberg (2003)
15. Li, C.: Describing and utilizing Constraints to Answer Queries in Data Integration Systems. In: IIWeb (2003)
16. Christiansen, H., Martinenghi, D.: Simplification of Integrity Constraints for Data Integration. In: Seipel, D., Turull-Torres, J.M.a. (eds.) FoIKS 2004. LNCS, vol. 2942, pp. 31–48. Springer, Heidelberg (2004)
17. Turker, C., Saake, G.: Consistent Handling of Integrity Constraints and Extensional Assertions for Schema Integration. In: Eder, J., Rozman, I., Welzer, T. (eds.) ADBIS 1999. LNCS, vol. 1691, pp. 31–45. Springer, Heidelberg (1999)
18. Cali, A., Calvanese, D., Giacomo, G.D., Lenzerini, M.: On the Expressive Power of Data Integration Systems. In: Spaccapietra, S., March, S.T., Kambayashi, Y. (eds.) ER 2002. LNCS, vol. 2503, pp. 338–350. Springer, Heidelberg (2002)

19. Cali, A., Calvanese, D., Giacomo, G.D., Lenzerini, M.: On the Role of Integrity Constraints in Data Integration. In: Bulletin of the IEEE Computer Society Technical Committee on Data Engineering (2002)

20. Fuxman, A., Miller, R.J.: Towards Inconsistency Management in Data Integration Systems. In: IIWeb (2003)

21. Popa, L., Velegrakis, Y., Miller, R.J., Hernandez, M.A., Fagin, R.: Translating the web data. In: VLDB, pp. 598–609 (2002)

22. Davidson, S., Fan, W., Hara, C., Qin, J.: Propagating XML Constraints to Relations. In: ICDE, pp. 543–554 (2003)

23. Lee, D., Chu, W.W.: Constraint Preserving Transformation from XML Document Type Definition to Relational Schema. In: Laender, A.H.F., Liddle, S.W., Storey, V.C. (eds.) ER 2000. LNCS, vol. 1920, pp. 323–338. Springer, Heidelberg (2000)

24. Liu, Y., Zhong, H., Wang, Y.: XML Constraints Preservation in Relational Schema. In: CEC-East (2004)

25. Wang, Q., Wu, H., Xiao, J., Zhou, A.: Deriving Relation Keys from XML Keys. In: ADC 2003 (2003)

26. Liu, Y., Zhong, H., Wang, Y.: Capturing XML Constraints with Relational Schema. In: CIT 2004 (2004)

27. Liu, C., Vincent, M., Liu, J.: Constraint Preserving Transformation from Relational schema to XML Schema. In: World Wide Web: Internet and Web Information Systems, vol. 9, pp. 93–110 (2006)

28. Deutsch, A., Tannen, V.: MARS: A System for Publishing XML from Mixed and Redundant Storage. In: VLDB (2003)

29. Bertino E., Ferrari E.: XML and Data Integration. In: IEEE internet computing, pp. 75-76(2001)

30. Benedikt, M., Chan, C.Y., Fan, W., Freire, J., Rastogi, R.: Capturing both Types and Constraints in Data Integration. In: Sigmod (2003)

31. Barbosa, D., Freire, J., Mendelzon, A.O.: Information Preservation in XML-to-Relational Mappings. In: Bellahsène, Z., Milo, T., Rys, M., Suciu, D., Unland, R. (eds.) XSym 2004. LNCS, vol. 3186, pp. 66–81. Springer, Heidelberg (2004)

32. Jiang, H., Ho, H., Popa, L., Han, W.: Mapping-Driven XML Transforamtion. In: WWW, pp. 1063–1072 (2007)

33. Fan, W.: XML Constraints: Specification, Analysis, and Applications. In: DEXA, pp. 805–809 (2005)

34. Fan, W., Simeon, J.: Integrity constraints for XML. In: PODS, pp. 23–34 (2000)

35. McBrien, P., Poulovassilis, A.: A formalisation of semantic schema integration. Information Systems 23, 307–334 (1998)

36. McBrien, P., Poulovassilis, A.: Schema Evolution in Heterogeneous Database Architectures,A Schema Transformation Approach. In: Pidduck, A.B., Mylopoulos, J., Woo, C.C., Ozsu, M.T. (eds.) CAiSE 2002. LNCS, vol. 2348, p. 484. Springer, Heidelberg (2002)

37. Poulovassilis, A., Brien, P.M.: A General Formal Framework for Schema Transformation. Data and Knowledge Engineering 28 (1998)

38. McBrien, P., Poulovassils, A.: Data Integration by Bi-Directional Schema Transformation Rules. In: ICDE, pp. 227–238 (2003)

39. Hartmann, S., Köhler, H., Link, S., Trinh, T., Wang, J.: On the Notion of an XML Key. In: Schewe, K.-D., Thalheim, B. (eds.) SDKB 2008. LNCS, vol. 4925, pp. 103–112. Springer, Heidelberg (2008)

Comparative Analysis of XLMiner and Weka for Association Rule Mining and Clustering

A.M. Khattak, A.M. Khan, Tahir Rasheed, Sungyoung Lee, and Young-Koo Lee

Department of Computer Engineering, Kyung Hee University, Korea
{asad.masood,kadil,tahir,sylee}@oslab.ac.kr, yklee@khu.ac.kr

Abstract. Retaining a customer is preferred more than attracting new customers. Business organizations are adopting different strategies to facilitate their customers in verity of ways, so that these customers keep on buying from them. Association Rule Mining (ARM) is one of the strategies that find out correspondence/association among the items sold together by applying basket analysis. The clustering technique is also used for different advantages like; recognizing class of most sold products, classifying customers based on their buying behavior and their power of purchase. Different researchers have provided different algorithms for both ARM and Clustering, and are implemented in different data mining tools. In this paper, we have compared the results of these algorithms against their implementation in Weka and XLMiner. For this comparison we have used the transaction data of Sales Day (a super store). The results are very encouraging and also produced valuable information for sales and business improvements.

Keywords: Association Rule Mining, Clustering, Weka, XLMiner.

1 Introduction

Now-a-days, retaining old customers is preferred more than attracting new customers. Business organizations are adopting different strategies to facilitate their customers in verity of different ways, so that these customers keep on buying from them. Association Rule Mining (ARM) [AIS93] is one of the strategies that have two fold advantages to the business organization after applying the basket analysis. 1) It helps customers to get all the related items from one place and that save their time from visiting different places of the store. 2) It helps organization in more selling of items by placing items closer that are sold together. Different business organizations around the world have used basket analysis technique; among these; *Wal Mart*[1] is the most famous example. Clustering technique is used for classifying data based on some of its characteristics into different classes that eventually help users/organizations to further smoothen their business process. Clustering results in different advantages for business organizations; 1) to recognize the class of most sold products, 2) classifying customers based on their buying behavior and their power of purchase, 3) classifying customers

---

[1] http://walmartstores.com/

D. Ślęzak et al. (Eds.): DTA 2009, CCIS 64, pp. 82–89, 2009.

arrivals in different time slots based on customers arrival time, and 4) identifying item(s) source for major trade.

Considering high dimensional data with noise and outliers, ARM and Clustering is a challenging task especially when data is very huge and complex. As discussed above, different researchers have provided different algorithms for ARM [LiH04] and Clustering [ESK03 and KaR90] that helps user to properly and efficiently achieve their objectives. The Apriori [LiH04] algorithm is used for ARM; it had a problem of candidate set generation. This problem was removed, so the new improved Apriori algorithm reduce the time of scanning candidate set. It uses the hash tree to store the candidate sets that facilitate in solving the frequent set counting problem and is now more optimized based on time factor. The same way K-Means [KaR90] algorithm is used for Clustering of data based on the parameter(s) specified.

In this paper, we have compared the results of two Data Mining tools i.e. XLMiner [XLMiner] and Weka [WiF05] for ARM and Clustering with Apriori and K-Means algorithms respectively. We have tested these algorithms of both the tools using daily transaction data from Point of Sale system of a super store *Sales Day (SD)*. We tested the Apriori and K-Means algorithms from both Weka and XLMiner on data of year 2007 of SD. By varying the parameters (i.e. support and confidence) for these algorithms; we got very interesting results for ARM discussed later. The same way, we also have tested both the tools with K-Mean algorithm for clustering of the data to identify the different classes of items sold of particular amount and users. The customers are clustered based on their buying power, time and power of buying, most frequent customers, and transaction with amount of transactions that helps in focused advertisement based on customer arrival time and their buying behavior.

Rest of the paper is organized as follows; Section 2 is related to preprocessing of Sales Day data to be used by the algorithms. Section 3 presents experimental results after applying both ARM and Clustering. In Section 4 is based on conclusions and future directions.

2 Data Normalization

Before applying the algorithms on data, first we need to normalize the transaction data for the algorithms to work on. The daily transaction data of Sales Day (SD) store as shown in Figure 1 is in organization required format, which is a high dimensional and complex data that is not useable by the algorithms. For this reason, we have first converted the data from the organization required format to the format required to be used for experimentation. Figure 1-A shows the schema of SD Point of Sale (PoS) system where the SD data is very much redundant as clear from Figure 1-B. For every item sold in a single transaction, there is a complete row for that item and repeating the same order data again and again. We have developed a MS-Visual Basic 6.0 application using MS-SQL 2000 queries to translate the data in to a single row (pivoting) for every ordered transaction as shown in Figure 1-C. We have worked on the transaction data of year 2007 and tested both the tools for ARM and Clustering. The transaction had a variability in number of items contained in them e.g., a person may buy only a milk or a snack pack (i.e. only one item) but a transaction may contain a whole variety of items that range from daily use to occasionally used items

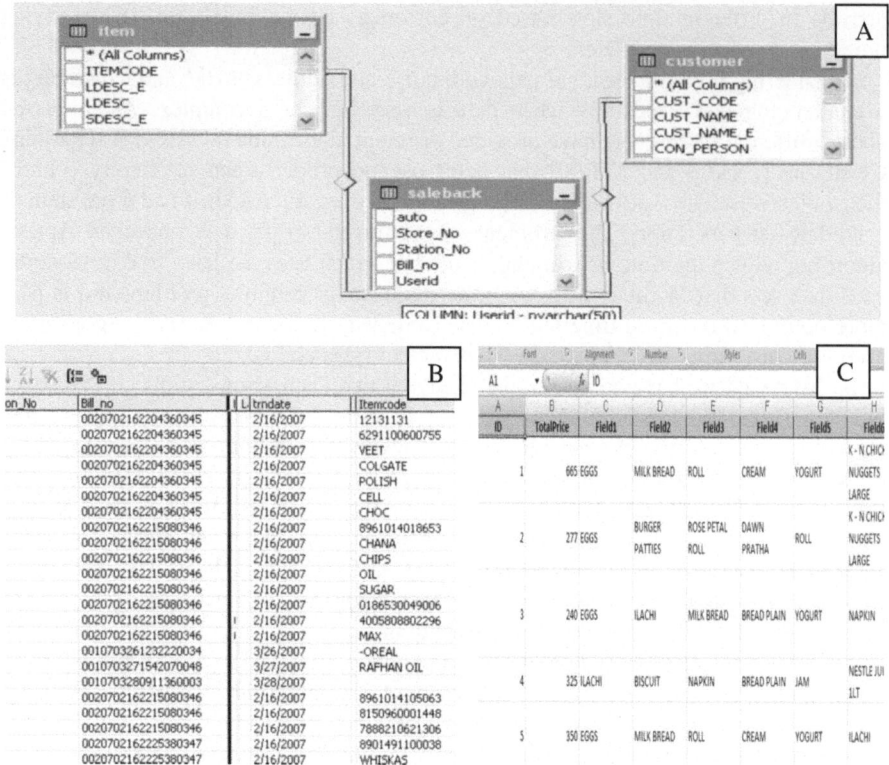

Fig. 1. (A) shows the original structure of transaction data storage in Sales Day Point of Sale system, (B) shows that for all the items sold in a transection is having a separate row entry, (C) shows the converted transections from multiple rows for one order to a single row.

that make the item set up to 60-80 products in it. We have fixed the number of items sold in a transaction to 12 items per transaction and any transaction having items more or less than 12 items are discarded.

3   Experimental Results

Here we discuss in detail the comparative study of both the tools (i.e. Weka and XLMiner) for ARM and Clustering results over the transaction data normalized in the previous section. We have divided this work into two sections where first one focus on ARM and the second is focusing on Clustering. In Section 3.1 we will discuss ARM using both Weka and XLMiner and after that in Section 3.2 we will discuss Clustering results. The results from both tools will be compared in these two subsections.

3.1   Association Rule Mining (ARM)

ARM finds interesting associations and/or correlation relationships among large set of data items. It infers attribute value conditions that occur frequently together in a given dataset e.g. Market Basket Analysis. Our goal is also to mine the Association Rules among data items from the transactions data of a super store Sales Day. The association rules provide information in the form of "if-then" statements where these are probabilistic in nature.

Among the different ARM algorithms available like; 1) Apriori, 2) Filtered Associations, 3) Predictive Apriori, and 4) Tertus, we choose to implement Apriori despite its multi scan drawback but the rules generated by Apriori are the most appropriate and finer granulized. To start working with Apriori for ARM, we have specified the environment variables as: 6994 instances of transactions with 12 attributes. The minimum support for ARM is set to 0.6 while minimum confidence is 0.9 with 20 numbers of cycles performed. Based on this input data for Apriori in both Weka and XLMiner, the association rules are mined (see Figure 2).

Results from both the tools depict same rules, while the representation of rules in these tools is different. For instance as shown in Figure 2-A, the confidence for CAKE to be purchased by the customer is 100 % if that customer is going to purchase BISCUIT, MILK, MILK BREAD, and NOODLE. The same rule is also represented in Figure 2-B of Weka where all the items are separately mapped with all the other items, and it also gives the confidence as 1 (which is equivalent to 100 %).

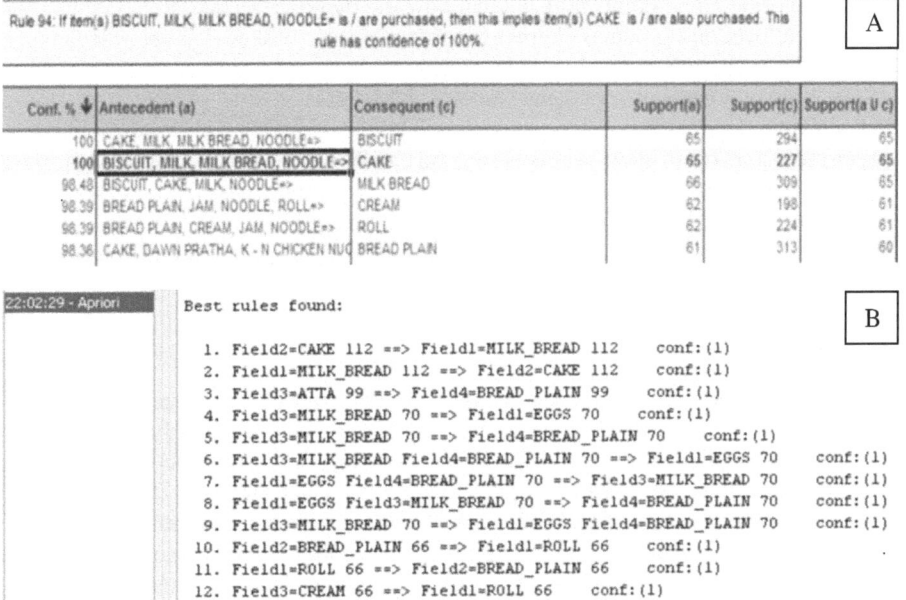

Fig. 2. (A) Shows the association rules identified using XLMiner. The yellow strip above display the complete rule selected in the table. (B) Shows the association rules mined using Weka.

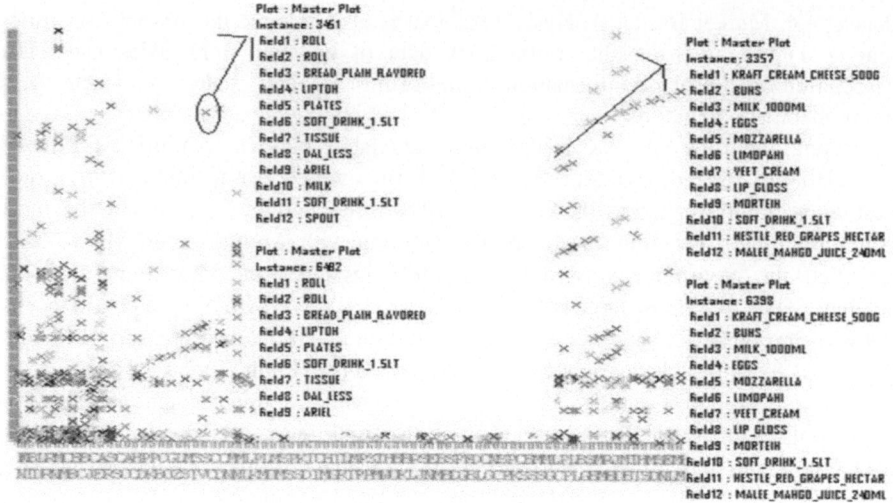

Fig. 3. Visualization of association rules using Weka

From the above Figure 3, it can be seen that how the associations are distributed over a plot. The circle points indicate that how the two transactions are related on the confidence level and products (items) occurring in their basket. This plot depicts the transactions with their respective basket items.

The associations/correspondence among sold items is one of the most useful source/results for the business organization. These associations are used by business organizations to resort their products in a way to place the most frequent sold items together. This also facilitates customers in quick checkout. One other strategy is to place the most sold items in different places. In this case the customer will have to visit different places in the store and will have a look at other different items available, that will increase their probability of been purchased by the customers.

3.2 Clustering

Clusters are often computed using a fast, heuristic method that generally produces good (but not necessarily optimal) solutions. The K-Means algorithm is one such method. We applied K-Means Clustering algorithm on the transaction data using XLMiner and Weka. In XLMiner, to do clustering, we enter the data range that needs to be processed and move the variables of interest to the selected variables box. Here, it is visible in Figure 4-B that the numbers of clusters are 4 and the attribute selected for the clustering is *TotalPrice* of transactions. It also represents the mean of clusters. Figure 4-A shows different transactions that are classified in different clusters, while Figure 4-C represents the objects (clusters) and the inter-cluster distance among them.

Weka uses the centroids positions for calculation of clusters. We have tested the data with K-Means using Weka and got 3 clusters as number of clusters depends on choice of initial centroids, choice of distance measure, and stopping criterion that we defined.

XLMiner : k-Means Clustering - Predicted Clusters   A

Row Id.	Cluster id	Dist clust-1	Dist clust-2	Dist clust-3	Dist clust-4	TotalPrice
1	3	467.38	966.33	0.245	3085	665
2	1	79.382	1354.3	388.24	3473	277
3	1	42.382	1391.3	425.24	3510	240
4	1	127.38	1306.3	340.24	3425	325
5	1	152.38	1281.3	315.24	3400	350
6	1	17.618	1451.3	485.24	3570	180
7	1	162.38	1271.3	305.24	3390	360
8	1	15.618	1449.3	483.24	3568	182
9	3	502.38	931.33	34.755	3050	700
10	1	42.618	1476.3	510.24	3595	155
11	3	287.38	1146.3	180.24	3265	485
12	1	2.618	1436.3	470.24	3555	195
13	2	1052.4	381.33	584.76	2500	1250
14	3	442.38	991.33	25.245	3110	640
15	1	102.38	1331.3	365.24	3450	300

B

Cluster	TotalPrice
Cluster-1	197.618
Cluster-2	1631.33
Cluster-3	665.245
Cluster-4	3750

Distance between cluster centers	Cluster-1	Cluster-2	Cluster-3	Cluster-4
Cluster-1	0	1433.712	467.627	3552.382
Cluster-2	1433.712	0	966.085	2118.67
Cluster-3	467.627	966.085	0	3084.755
Cluster-4	3552.382	2118.67	3084.755	0

C

Fig. 4. (A) Shows classification of transactions in clusters, (B) shows number of clusters and their distance, and (C) visually represent inter-cluster distance

These are overlapping clusters that are obtained by the range of the mean and standard deviation specified. By analysis of the visual plot obtained and shown in Figure 5. It is clear from Figure 5 that the clusters are roughly distributed. The color distribution for clusters is; Blue: Cluster 0, Green: Cluster 1, and Red = Cluster 2. The circle points show that at that point which cluster value it is.

3.3 Analysis of Results

The ARM and Clustering work conducted here in this paper is basically for the purpose of comparative analysis of Weka and XLMiner with Apriori and K-Means algorithms. We have tested both the tools and for ARM they gave exact answers but during different experiments we performed for Clustering generated different clusters. Beside this comparison objective, we also got some very interesting results based on the transactions data. We got the most frequent sold items in a basket that helped the organization in re-organizing the sale strategy for these items to improve their sale. As shown above, in Clustering we have also classified the transactions data on different parameters like; age group of customers, age group plus purchase power, most sold items, time of high customer traffic, and time of high purchase power

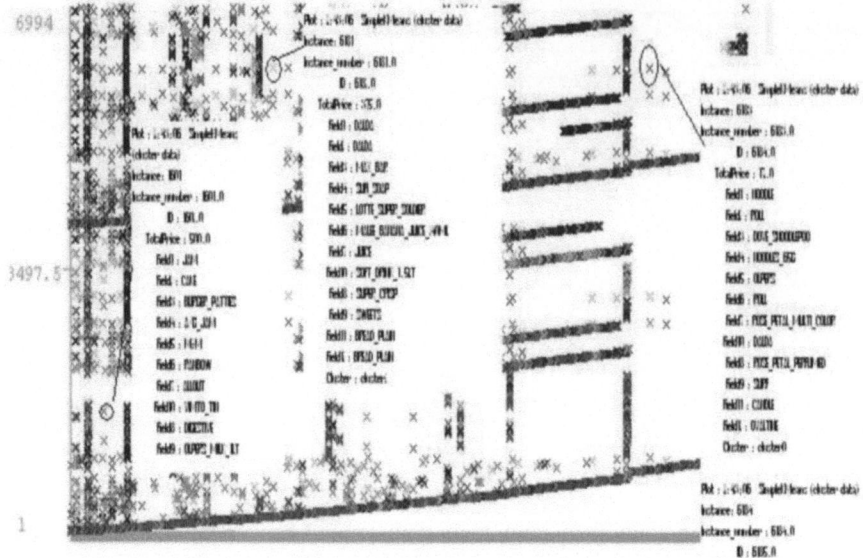

Fig. 5. Visualization of clustered transactions using Weka

customers arrival. These classified (Clustered) transactions are used by the business organization to great advantage. These results are very useful, for instance; with the help of cluster i.e. Time of high purchase power customer arrival, the organization can lunch new high cost products for these customers. Knowing the arrival time of particular type of customers can also be used for focused advertisement of product of interest to the arriving customers. These results also avoid the out-of-stock situation as it gives information about most sold items.

4   Conclusions and Step Ahead

Association Rule Mining and Clustering is a well established area of data mining. These are used for extracting some hidden information from a huge repository of raw data. In this paper we used these two techniques with their Apriori and K-Means algorithms implemented in Weka and XLMiner to analyze the trend of sale at a super store Sales Day. We have compared these algorithms by using Weka and XLMiner over Sales Day data and got very encouraging results that not only satisfy the implementation of these algorithms in both the tools as same but also support the business organization for customer support and future extension in their business. We are planning to extend our work to different tools for more algorithms and use the results to business advantages.

Acknowledgement

This research was supported by the MKE (Ministry of Knowledge Economy), Korea, under the ITRC (Information Technology Research Center) support program supervised by the IITA (Institute of Information Technology Advancement)" (IITA-2009-(C1090-0902-0002)) and was supported by the IT R&D program of MKE/KEIT, [10032105, Development of Realistic Multiverse Game Engine Technology].

This work also was supported by the Brain Korea 21 projects and Korea Science & Engineering Foundation (KOSEF) grant funded by the Korea government(MOST) (No. 2008-1342).

References

[AIS93] Agrawal, R., Imielinski, T., Swami, A.: Mining Associations between seta of items in Massive Databases. In: proc. Of the ACM-SIGMOD 1993 int'l conf. on Management of Data, Washingtom D.c USA, pp. 207–216 (1993)

[ESK03] Ertoz, L., Steinbach, M., Kumar, V.: Finding Clusters of Different Sizes, Shapes, and Densities in Noisy. In: SIAM International Conference on Data Mining, (Feburary 20, 2003)

[KaR90] Kaufman, L., Rousseeuw, P.J.: Finding Groups in Data: An Introduction to Cluster Analysis. Wiley Series in Probability and Statistics. John Wiley and Sons, New York (1990)

[LiH04] Liu, X.W., He, P.L.: The research of improved association rules mining Apriori algorithm. In: Proceedings of 2004 International Conference on Machine Learning and Cybernetics, August 26-29, vol. 3, pp. 1577–1579 (2004)

[WiF05] Witten, I.H., Frank, E.: Data Mining: Practical machine learning tools and techniques, 2nd edn. Morgan Kaufmann, San Francisco (2005)

[XLMiner] Data Mining Add-In For Excel, http://www.resample.com/xlminer/

Infobright for Analyzing Social Sciences Data

Julia Ann Johnson[1] and Genevieve Marie Johnson[2]

[1] Department of Mathematics and Computer Science, Laurentian University,
Sudbury, Ontario, Canada P3E 2C6
jjohnson@cs.laurentian.ca
[2] Department of Psychology, Grant MacEwan University,
Edmonton, Alberta, Canada T5J 4S2
johnsong@macewan.ca

Abstract. There are considerable challenges in analyzing, interpreting, and reporting word-based social sciences data. Infobright data warehousing technology was used to analyze a typical data set from the social sciences. Infobright was found to require augmentation for analyzing qualitative data provided as short stories by human subjects. A requirements specification for mining data that are subject to interpretation is proposed and left to the Infobright designers to implement should they so choose. Infobright was chosen as a system for implementing the data set because its rough set based intelligence appears to be extensible with moderate effort to implement the data warehousing requirements of word based data.

Keywords: Rough set, data warehousing, Infobright, semantic similarity, qualitative data, qualitative data analysis, knowledge grid.

1 Introduction

Infobright (IB) is a database server that uses a rough set based data compression method to efficiently process queries on very large databases. The use of information that is already provided by IB or can be easily provided with minor modification is advanced for analysis of data from the social sciences.

IB is reviewed in Section 2. A typical social sciences data set is reviewed in Section 3. Implementation of that data set using IB is described in Section 4. Extensions to IB for better analysis of such data are proposed in Section 5. Methodology for implementing the proposed extensions while streamlining with existing IB methodology is presented in Section 6. Conclusions are provided in Section 7.

2 Infobright

Historically, with conventional databases it has been possible to answer, for example, count queries by accessing only the index structure allowing such queries to be executed efficiently. For very large databases and more complex queries such a

D. Ślęzak et al. (Eds.): DTA 2009, CCIS 64, pp. 90–98, 2009.
© Springer-Verlag Berlin Heidelberg 2009

strategy breaks down and must be augmented with newer technologies for efficient query processing.

Infobright [5][6] uses a column oriented database architecture to improve efficiency of querying large databases. Column oriented databases as opposed to row oriented ones lend themselves to data compression techniques. Since all values in a column are of the same type, the values of each column may be split into separately compressed value chunks. Information about the column type and the patterns occurring inside the value chunks characterize the column. The amount of information stored to describe the value chunks tends to be far smaller than that required to represent the row chunks of comparable size.

Infobright uses what is referred to as the database knowledge grid that equates to metadata of conventional databases but organized with knowledge nodes of compact information about the value chunks. The historical counterpart in conventional databases is data block indexing, but now the blocks correspond to large portions of the database. Much of query processing can be accomplished using the knowledge grid without need or with significantly limited need to fetch compressed data. We would like to see the knowledge grid of IB used not only for achieving goals of physical storage, but also, for providing information at the database conceptual level.

3 Social Sciences Data

The social sciences comprise academic disciplines concerned with the study of the social life of human groups and individuals including anthropology, communication studies, cultural studies, demography, economics, education, political science, psychology, social work, and sociology. Inevitably, research methods in the social sciences include collecting data on human subjects. Research tools for collecting social sciences data include direct observation in various contexts, questionnaires completed by individuals, and the administration of standardized instruments such as IQ tests. Two types of data emerge from such collection strategies, -- qualitative and quantitative. Generally speaking, qualitative data are expressed in words, texts, narratives, pictures, and/or observations; quantitative data reflect numerical representation of the world.

Conventionally, word-based data are manually coded in order to organize a very large amount of words into manageable chunks. Such subjective treatment of data introduces bias and error as different individuals interpret text differently [1]. An example may clarify some of the challenges in qualitative social sciences data analysis.

3.1 Example of Social Sciences Data: Virtual Communities for Those Who Self-injure

Non-suicidal self-injury (SI) is defined as direct, deliberate destruction of one's own body tissue without suicidal intent. Polk and Liss [4] found that 20% of college students reported having self-injured at some point in their lives. Many self-injurers find support in virtual communities that typically include a website with e-message boards. Important information for psychologists and e-health practitioners includes description of individuals who participate in virtual communities for those who SI (e.g., the nature of self-harm, reasons for participating in virtual communities, and perception of the effect of such participation on level of SI).

Individuals who participate in virtual communities responded to questions posted on two e-message boards for those who SI. Questions were all open-ended and thus only answerable with text. For purposes of illustration, Table 1 provides some of the posted questions and the responses of one participant. Note that in qualitative data, grammar, spelling, and punctuation are not corrected.

Typical responses contain many words and are difficult to precisely categorize. In general, the greater the number of words in a response, the greater the time required and the greater the influence of subjective interpretation. To confirm, two trained research assistants independently coded the data. In both cases, coding took weeks of focused effort. Divergent interpretation of data was apparent between the two coders and, in both cases, description of methods of SI were limited, despite considerable effort and expenditure of personnel resources. There would be no need for coding methods if queries could be evaluated based on semantic comparison of a text field with a textual column value.

3.2 Semantic Similarity Comparison Metrics

Algorithms for the various aspects of measuring similarity are available that rely on use of the WordNet lexical database. Wordnet gives specific meanings of words and establishes connections between parts of speech. An interface is available that accepts words and gives a measure of their similarity. Various algorithms for semantic similarity have been developed by researchers including Resnik, Lin, Jiang-Conrath, Leacock-Chodorow, Hirst-St.Onge, Wu-Palmer, Banerjee-Pedersen, and Patwardhan-Pedersen. The software for WordNet is available as open source, and the algorithms are available for example as a Perl module or Java pseudo code. The WordNet route

Table 1. Posted Questions and Sample Response (Abbreviated)

Posted Questions	Sample Response
Why and how do you self-injure?	... I discovered it after a bad day when I was 5 and had broken a glass and accidently cut myself with the glass...
How long have you self-injured?	I started around 5 or 6 and I am 34 now.
Have you ever tried, or are you now trying, to stop self-injuring?	I have never really made a major effort to quit but there have been periods where I just stopped for extended periods ...
If you are now attempting to stop self-injuring ... what methods ...	One of the methods that I used to use was to drink which nearly always backfired. Now days I try to write ...
Do you have an eating disorder, or abuse drugs or alcohol?	I do not have an eating disorder but as a form of control when I try to stop cutting I have had disordered eating ...
Do you feel you have control ...	For the most part yes
When start using these boards?	I joined ... a little over 3 years ago.
Why do you use this board?	I have been visiting as a regular member ...
How often do you visit ...	I have found that it gives me an emotional outlet, allows me the chance to say what I will in a nonjudgemental manner ...

has been excluded from further consideration because grammatical correctness and spelling accuracy cannot be assumed for social sciences qualitative data. However, WordNet has been used to detect spelling errors that go unnoticed by a regular spelling checker (Hirst and Budanitsky 2005).

Chien & Immorlica [2] investigate the idea of discovering semantically similar queries of search engines. Their technique rests on the similarity in behavior of the queries over time. They developed a method of finding temporally correlated input queries to serve as a means of quantifying the relatedness between queries. Their work may be relevant for relatedness of questions, but logically related qualitative responses may not be temporally correlated as are the requests of a web server. However, this work may be applicable to social sciences data for measuring semantic similarity of responses collected on an ongoing basis. In the investigation at hand, the focus has been on analysis of responses that have been collected from a web site posted for a short period of time.

3.3 Existing Software for Analysis of Social Sciences Data

To some extent, the coding of qualitative social sciences data has been improved with the development of software. Current code based software includes content analysis tools, word frequencies, word indexing with key word in context retrieval, and text based searching tools [3]. Such software systems are, however, inadequate for making meaning of text. They amount to data management systems requiring the user to reformulate the data in a preprocessing step. To avoid the subjectivity so introduced a view based on rough sets was investigated.

4 Infobright Implementation

IB uses the concept of a rough set to determine for a given query which stored data packs are irrelevant (disjoint with the answer set), relevant (fully inside the set) and suspect (overlapping with the answer). Only the suspect data packs need to be decompressed because for those it is necessary to determine exactly what parts of their data satisfy the query at hand.

The use of such data compression techniques is needed for efficiently storing and retrieving information in large stores of word-based qualitative data. Compression is even more important for ongoing collection of qualitative data. But, information generated as a result of decompression may also find use in evaluation of the meaning of queries.

IB returns statistics about the time required to evaluate a query. There is likely a correlation between the amount of time required to evaluate a query and the amount of exact computation required for the query. Exact computation occurs when data packs must be decompressed. A way to test the hypothesis that information generated as a result of decompression has semantic utility is to see if there is a correlation between query speed and semantic relatedness of queries.

4.1 Database Design

In qualitative data, column values that tell a story are of interest. Column names are likewise one or more short sentences, usually questions. In the selected design each column name corresponds with a question, and there is exactly one record for each subject. A column extension corresponds to the answers by all subjects to the question given by that column.

Experience was gained on suitable column names when Excel tables were used to record coded data. The tables were designed with abbreviated column names that expanded to the full text of the question. In Microsoft Excel, columns themselves may be stretched having the effect of revealing previously hidden parts of long column names. In similar manner when using IB, it would be helpful to be able to point to a column entry and have it expand to the full text comprising the column value. Otherwise, a meaningful tabular display of qualitative data to fit conventional display media would not be possible.

4.2 Database Definition and Loading

Infobright Enterprise Edition was obtained on the basis of a special academic promotional offer. It was installed on an 8-core, 8 gig ram server running the Debian operating system (a flavor of Linux). IB's MySQL pluggable storage engine architecture allowed the database server to be accessed using an SQL client running on Windows.

A database schema was defined and implemented on the IB server (Fig. 1) with 12 columns (questions). The database was populated with 67 rows corresponding to answers given by as many subjects.

Abbreviation of column names was required to fit the MySQL constraints on names. Take for example Q6 that reads "If you are now attempting to stop self-injuring, or have previously tried to do so, what methods do you use when you feel the urge to self-injure? Where did you learn these methods?" The text of such questions is too long for a name.

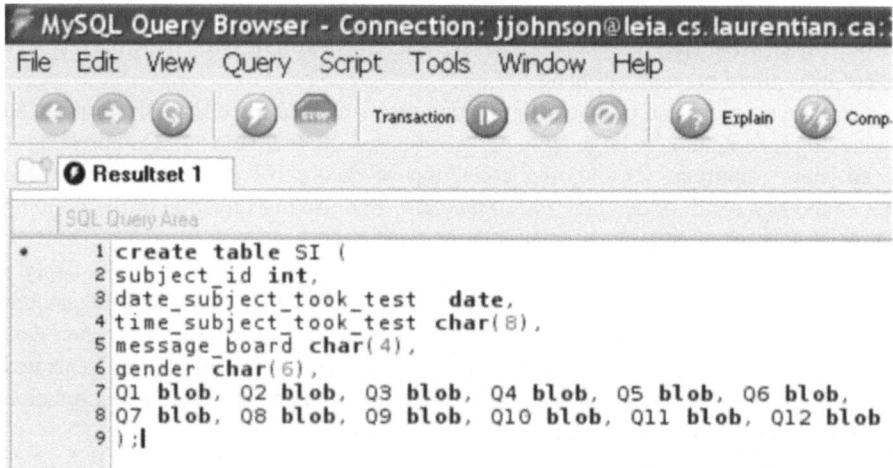

Fig. 1. Schema Definition using MySQL query browser connected to Infobright server

Because IB designers have focused on storing large amounts of data, keys are not supported. The command in its simplest form for loading data into a table follows:

LOAD DATA INFILE '/c:/tilde_data'
INTO TABLE SI
[FIELDS
[TERMINATED BY '~']
];

/c:/tilde_data refers to the file in which the data are located. The tilde as column entry delimiter was arbitrarily selected. Options are available for handling delimiters that also appear within column entries. If tildes were expected to occur in responses given by subjects, those options would have been employed.

The data required preprocessing to permit use of the IB LOAD command that was specifically designed for large quantities of data. The syntactic requirements of the MySQL INSERT statement were to be avoided. The preliminary field values and answers given to the first few questions for subject 001 in preprocessed form follows:

~001~   Nov. 13~   ~9:37pm~   ~bus ~
~ no answer From question 4, the answer can be supposed to be '34' ~
~ Female ~
~ I started cutting completely by accident. I discovered it after a bad day when I was 5 and had broken a glass and accidently cut myself with the glass. I noticed that it helped release the intense emotions and made me feel more able to breathe. There were times when I would burn but that was an entirely different sensation than cutting and was rare.~
~ I started around 5 or 6 and I am 34 now.~

Recall the table SI (self injury) as illustrated in Fig.2 into which these data were loaded.

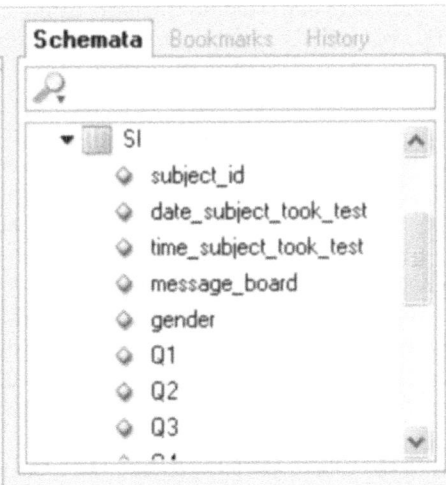

Fig. 2. Fields of self injury table to collect preliminary information and answers to questions. Although subject_id uniquely identifies subjects, we have no facility for specifying a key constraint.

4.3 Database Query

Specific requests on the data were formulated by guessing about likely relationships eg.

1. Do older individuals use different methods of SI than younger individuals?
2. What is different between individuals who say their SI has increased since visiting the boards and those who say their SI has decreased?
3. What differentiates cutters who have eating disorders (or substance abuse issues) and those who do not?
4. Do boys self injure for the same reasons as girls?

Let *reason* denote the "why" part of Q1 (Why and how do you self-injure?) A MySQL query to partially answer request 4 looks like the following:

```
SELECT M.subject, F.subject FROM SI M, SI F
WHERE M.reason = F.reason AND M.gender ='M' and F.gender ='F';
```

As of now, there is no native support for weighing the semantic equivalence of two sentences either through proprietary measures or by any SQL specific extensions (Quoted from the IB Help desk). Therefore, the above query returns an empty (null) answer even if the semantic content of the two operands is identical (M.reason = F.reason). There were opportunities to leverage IB technology to help with the problem by sharing in the Infobright Forums.

5 Proposed Extensions to Infobright

Several possibilities for introducing semantic similarity were studied: 1) augmentation of query results to get in effect semantic sentence or paragraph comparison 2) native support in IB for weighing the semantic equivalence of two sentences, 3) augmentation of the database in a preprocessing step to include columns that contain coded information that would permit a KDD approach to computing semantic similarity.

A strategy based on possibility 3) was discussed in the IB forums. The idea proposed was to handle semantic similarity at the data preprocessing level, comparable to the ETL (Extract Transform Load) process required to convert an existing database to an IB one. Similarity was to be expressed by means of additional columns in the data table created prior to loading the data. However, this strategy is fraught with the same problems that were outlined in Section 3.1 concerning subjective hand coding of the meaning of qualitative data. Two coders will invariable provide two different interpretations of the same data.

For option 1), given appropriate additional columns, the semantic similarity should be at least roughly expressible by means of SQL conditions. Algorithms for semantic similarity may then be at least partially adapted to a MySQL query interface. Again, the additional columns introduce subjectivity and considerable expenditure of personnel resources.

Proposed extensions to IB follow: 1.) The provision of a means of associating with the results of relational expressions (>, =, IN, etc.) in queries an indication of whether the relationship is irrelevant, suspect or relevant 2.) In the case of an exact computation (suspect data packs), a measure of the amount of overlap. Such a measure differs from early measures of certainty and coverage associated with predictive rules generated by rough set based inductive learning algorithms. There the metrics were based on rows while here they are based on columns 3.) For the HAVING and GROUP BY clauses in queries a column oriented feature extraction and classification process that clusters texts within a given column by placing those requiring a high degree of exact computation (semantically similar ones) all in one cluster. 4.) Expansion of the statistics about query speed to include other information available during selective decompression to provide a more accurate indicator of the degree of overlap.

Additionally, we would like to see a facility provided whereby an IB database may be overlaid by a MySQL schema for specification and enforcement of key constraints and referential integrity constraints.

6   Streamlining Extensions with Existing Infobright Methodology

Implementation of the proposed extensions is expected to take advantage of the organization of an Infobright database as a rough level and an exact level. The rough level known as the knowledge grid with its "small, efficient, calculable units of information about data packs" reduces the need for decompression. (quoted from presentation made by Slezak at Japan and Korean conferences, 2009). It is anticipated that the knowledge grid could also be used to provide semantic information about whether a given text and one selected from a column are identical in meaning, not similar in meaning at all, or overlapping in meaning. An indicator of the degree of overlap is already partially provided by the statistics returned from IB upon each query evaluation.

7   Summary and Conclusions

For qualitative social sciences data, compression and selective decompression are needed due to the complexity and data concentration of both column entries and column names. Extensions to Infobright functionality have been proposed to allow it to be used in social sciences research. IB was investigated for its ability to reduce subjectivity of interpretation and alleviate the effort required for preparing qualitative data for analysis. Proposed minimal extensions include:

1.   Support for textual column values that tell a story
2.   The ability to distinguish texts from a given column that require no decompression from those in the same column that require some.
3.   Support for the query language to cluster texts from a given column based on the degree of decompression required to materialize them.
4.   Provision of additional parameters regarding the amount of decompression done.
5.   Expandable column names to reveal the full text of questions and expandable column entries to reveal the full text of answers given by respondents.

With this we have defined the kind of semantic similarity that will aid analysis of social sciences qualitative data. The information requested in items 1 through 4 is already available at the physical (storage) level for fast response to queries but we see it as also having a purpose at the conceptual level, specifically, for measuring the degree of overlap in the meaning of answers given by different subjects to corresponding questions.

The primary use of IB at Laurentian University is to teach database concepts to computer science students. We are in the process of developing the infrastructure for the undergraduate database course to augment the presentation of MySQL with hands on use of ICE (Infobright Corporate Edition). Data integrity achieved by means of constraints defined by database designers and enforced by the database management system must be covered at the undergraduate level. Hence, the possibility of overlaying a database (conveniently loaded using the simple IB LOAD command) with a schema of a richer structure and meaning was proposed. For our graduate knowledge base course to be offered next term a final project implementing a database of interest using IB will be expected. Infobright Enterprise Edition (IEE) running on our 8-core server with 8 gigs of RAM supports a maximum of 8 concurrent users sufficient for expected enrollments in a graduate course. Among the topics to be explored in the knowledge base course is the problem of how to achieve the minimal extensions to IB that were outlined in this paper.

References

1. Auerbach, C.F., Silverstein, B.L.: Qualitative data: An introduction to coding and analysis. New York University Press, New York (2003)
2. Chien, S., Immorlica, N.: Semantic similarity between search engine queries using temporal correlation. In: Int. World Wide Web Conf. WWW 2005, Chiba, Japan (2005)
3. Lewins, A., Silver, C.: Using software in qualitative research: A step-by-step guide. Sage, London (2007)
4. Polk, E., Liss, M.: Psychological characteristics of self-injurious behavior. Personality and Individual Differences 43, 567–577 (2007)
5. Ślęzak, D., Wróblewski, J., Eastwood, V., Synak, P.: Bright-house: An Analytic Data Warehouse for Ad-hoc Queries. PVLDB 1(2), 1337–1345 (2008)
6. Ślęzak, D., Wróblewski, J., Eastwood, V., Synak, P.: Rough Sets in Data Warehousing. In: Chan, C.-C., Grzymala-Busse, J.W., Ziarko, W.P. (eds.) RSCTC 2008. LNCS (LNAI), vol. 5306, pp. 505–507. Springer, Heidelberg (2008)

Enhanced Statistics
for Element-Centered XML Summaries

José de Aguiar Moraes Filho, Theo Härder, and Caetano Sauer

University of Kaiserslautern
P.O. Box 3049, D-67653, Kaiserslautern, Germany
{aguiar,haerder,csauer}@cs.uni-kl.de

Abstract. Element-centered XML summaries collect statistical information for document nodes and their axes relationships and aggregate them separately for each distinct element/attribute name. They have already partially proven their superiority in quality, space consumption, and evaluation performance. This kind of inversion seems to have more service capability than conventional approaches. Therefore, we refined and extended element-centered XML summaries to capture more statistical information and propose new estimation methods. We tested our ideas on a set of documents with largely varying characteristics.

1 Introduction

In recent years, many publications [1,2,3,4,5,6,7] proposed summarization methods providing statistics needed for XQuery/XPath optimization. Supporting only small and incomplete subsets of the rich XPath relationships, these methods often fail to cover all estimation requests (for up to 8 major axes relationships) of an optimizer. Furthermore, they widely differ in estimation accuracy delivered, storage space occupied, and memory footprint needed. Altogether, these deficiencies may heavily influence the process of query optimization. Therefore, no commonly agreed-upon solution is available so far.

We have developed an element-centered summary called EXsum [8] that focuses on the set of distinct element/attribute names of an XML document, putting aside the strict hierarchy between them. Instead of summarizing over the entire tree structure at a time and to keep track of (root-to-leaf) paths in the document, this method gathers node by node structural information for every distinct element name in the document tree. Compared to competitor approaches, this novel way to summarize XML documents is more expressive and outperforms them in many optimization scenarios [8].

The base version of EXsum can be made extensible to enhance statistical information to be recorded on an XML document. Therefore, we have developed a suitable extension of EXsum which contributes to the state of the art. This extension captures more information from XML documents, especially the fan-in and fan-out of the axis relationships controlled, enabling new estimation procedures, especially in case of path expressions involving more than two steps or expressions with predicates. An empirical cross-comparison with competitor algorithms is provided based on a set of well-known XML documents.

D. Ślęzak et al. (Eds.): DTA 2009, CCIS 64, pp. 99–106, 2009.

2   Extending EXsum

The original EXsum only captures the axis-related fan-out of all document nodes in a specific format ([OC]) by using a single counter per axis for each distinct element name. This axis-specific summarization of statistical data is called an ASPE spoke—*Axes Summary Per Element*. Computing fan-in and fan-out for every axis relationship may give us more opportunities to explore refined estimation methods. For this reason, we double

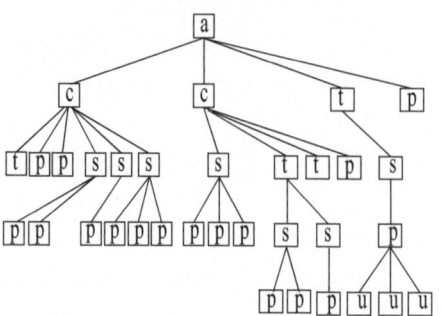

Fig. 1. A sample XML document

the number of counters: IC counters for the fan-in and OC counters for the fan-out ((([IC, OC])).

As a second observation, additional information called DPC (*Distinct Path Count*) is helpful to support some special estimation procedures. DPC counts the number of distinct path instances that satisfy a specific relationship, starting from the document root. In other words, if we have a relationship $s \rightarrow p$, we record the number of distinct rooted paths leading to s nodes involved in such a relationship. The tuples ([IC, OC]) stored for each element in the child and descendant spokes is upgraded to ([IC, OC, DPC]) to encompass DPC count.

Building Algorithm. To correctly count the node occurrences for EXsum, the plain building algorithm described in [8] must be modified. As for the plain EXsum format, the counting of axes occurrences is done for each element using a stack S. Counter calculation is straightforward for forward axes (descendant and child): we simply add 1 to the respective counter in the corresponding ASPE node, every time we find a descendant/child element in the stack.

Relationship counting in reverse axes (parent and ancestor) is, however, a bit more complex. We use an auxiliary list called *Element in Reverse Axis* (ERA). It maintains, for each element x in S, a list of all distinct nodes that were pushed onto S after x or, in other words, all distinct nodes under the subtree rooted by x. This means that, every time an element is pushed onto S, the list of each distinct element name currently in S is updated. Another use of ERA lists is to update IC/OC counters in every ASPE node involved in the computation. We exemplify the extended EXsum building process using the document in Fig. 1a. Furthermore, Fig. 2 illustrates the initial building steps of EXsum.

When the document root is visited, its name a is pushed onto S. In addition, ASPE(a) is allocated and all axes information that can be evaluated in this situation is recorded. In this case, we add 1 in ASPE(a) as the current number of a occurrences in the document and allocate an (empty) ERA list for a, currently the Top Of Stack (*TOS*) (Fig. 2a). In the next step, proceeding in document order, an element name c is located and pushed onto S. To control the allocation of ERA lists, we check if node y is the first occurrence under the subtree rooted

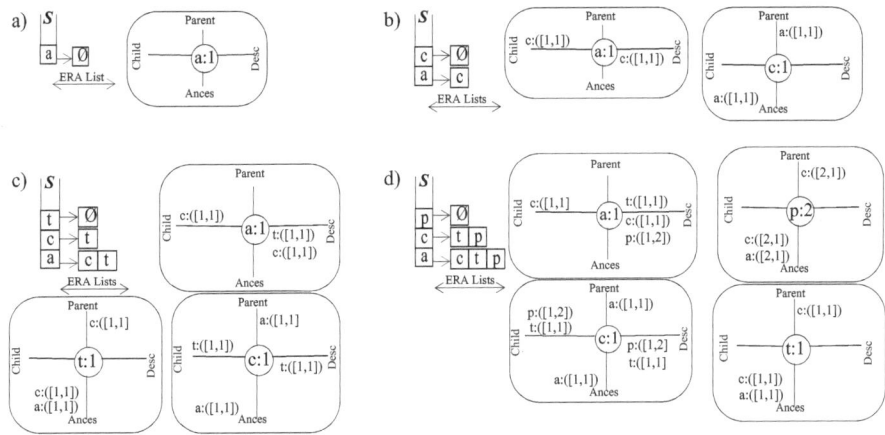

Fig. 2. States of EXsum and stack S for the initial building steps

by node x. To perform this check, we must look for node y in the ERA list of x. If no occurrence is found, we register y and return $True$. Because ASPE(c) is not present, it is created and the related axes information is added to a and c as follows. The IC/OC counters are adjusted in ASPE(a) and in ASPE(c). As it is the first time that an element c appears under (a subtree rooted by) a, the check returns $True$ and we include c in the ERA list of a. Thus, a c with ([1,1]) is included in the child spoke of ASPE(a). Accordingly, ASPE(c) has an a with ([1,1]) in the parent spoke indicating only one c and one a a in this subtree.

Additionally, we add a c with ([1,1]) in the descendant spoke of ASPE(a) and an a with ([1,1]) in the ancestor spoke of ASPE(c) (Fig. 2b). The main reason to do so is to be compliant with the axis definitions in the XPath specification. Hence, EXsum counts child (parent) and descendant (ancestor) relationships together in the descendant (ancestor) spoke and separately inserts child (parent) relationships only in the child (parent) spoke. Continuing the document traversal, a node with element name t is now visited ($S = [a, c, t]$) (Fig. 2c). Again, t is pushed onto S, ASPE(t) is created, and the axes information for t and its path elements c and a is completed. ERA lists of a and c now include a t and, again, both lists report that it is the first t encountered. Thus, an a with ([1,1]) appears in the ancestor spoke of ASPE(t). The same t-counters exist for the child spoke of ASPE(c), parent and ancestor spokes of ASPE(t), and for the descendant spoke of ASPE(a). As t has no children, t is popped out from S. Then, reaching the fourth element p, S and the counters are adjusted in the parent and ancestor spokes of ASPE(p), child and parent spokes of ASPE(c), and descendant spoke of ASPE(a). The states of EXsum and stack S at this point are shown in Fig. 2d.

The correct counting of elements in the reverse axes is highlighted when the process visits the fifth element (the second p, $S = [a, c, p]$). Here, ASPE(p) is already allocated and we have p in the ERA lists of c and a. Thus, it is not the first occurrence of p under the subtrees rooted by c and a. Therefore, we add 1 for ASPE(p) that now counts 2 and add also 1 for p in the child spoke of ASPE(c) and in the descendant spoke of ASPE(a) which now contain ([1,2]). Conversely, we do not add 1 for the OC counters of a and c in parent and ancestor spokes of ASPE(p), i.e, they keep ([2,1]). This mirrors the document structure in which there is one c as parent of two p nodes and, consequently, one a as ancestor of two p nodes. Hence, after a subtree is entirely traversed, we have obtained the correct values of the corresponding IC/OC counters.

Calculating DPC. To compute DPC, we need to maintain the set of all distinct rooted paths for each relationship. We have designed a procedure which processes every rooted path occurrence. For every given pair of related nodes, we maintain two sets, one for the child (child set) and the other for the descendant (descendant set) relationship.

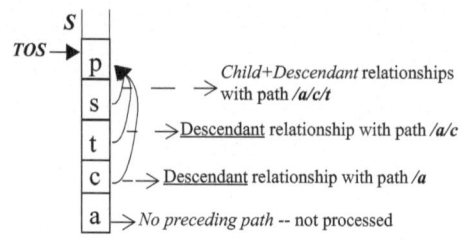

Fig. 3. Computing DPC

To explain how this it works, we take a practical example. Consider the path (a, c, t, s, p) in the document in Fig. 1. Fig. 3 illustrates which relationships and paths are computed. The TOS element in this case is p. The procedure starts by assigning the size of the path to n, which is 4 in this case. Then, we check for the value of n. The procedure is only executed for values of n greater than 2, because a path with 2 nodes contains only one child relationship and, therefore, no preceding distinct paths. Then, for every node i in the path before the TOS, we add an occurrence of the sub-path that leads from root to the descendant relationship between i and the TOS. For the particular case of the relationship between TOS and the element right before it, the path is also added to the child set. So, in the given example, the procedure starts with element c (position 2). Then, we take the descendant set of the relationship from c to p, denoted as a pair $(c; p)$, and add an occurrence of the path $/a$. The child set will be left untouched, as c is not at position TOS-1. Because sets are used, no duplicate elements will be added, and only distinct paths will populate them.

When going to the t element, the path to be added is $/a/c$. Reaching the fourth s, we need to add the path $/a/c/t$ to the relationship $(s; p)$. Moreover, it will also be added to the child set, as the element is positioned right before the TOS p. After completion of the document scan, EXsum building is finished; the resulting structure, including DPC counters, is shown in Fig. 4.

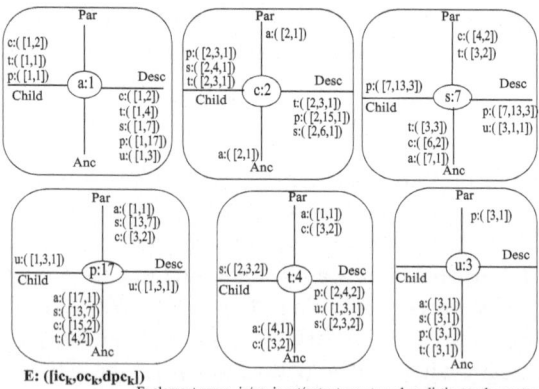

Fig. 4. EXsum for our sample document

Estimation Procedures. Because exact calculations cannot be performed by EXsum when n-step (n>2) path expressions come into play, [8] proposed the use of *interpolation* as an estimation procedure. Based on EXsum extensibility, we realize new procedures to compensate the decrease of accuracy in these cases. For the following explanations, we refer to the EXsum structure depicted in Fig. 4.

DPC Division. The DPC division procedure relies on the *uniform distribution assumption* of document paths leading to a location *step* captured by the DPC counter in the EXsum structure. The idea is to divide the $occ(step)$ cardinality by the related count.

Consider a path expression $/a/c/s/p$. To estimate step $/s$, $occ(/s)$ is given as follows. In ASPE(c), we search the child spoke for an s and find the OC and DPC counters. The estimation of $occ(/s) = OCcounter/DPCcounter$ delivers 4/1=4 as step estimation. This means that there is only one path leading to $c \rightarrow s$. For the next step $(/p)$, we find three distinct paths reaching $s \rightarrow p$: (a,c), (a,t) and (a,c,t). With an OC counter value 13 of p in the child spoke of ASPE(s), $occ(/p) = 13/3 = 4.3$, which is the estimated cardinality of the expression. DPC is also available for descendant steps. DPC for the $//s//p$ would deliver 3 $((a,c),$ (a,t) and $(a,c,t))$, because it corresponds to the number of paths leading to s nodes which have at least one p in its subtree. Thus, the estimate for $//s//p$ is $occ(//s//p) = 7/3$.

Previous Step Cardinality Division. The *Prev.Step* method uses the $occ(currentstep)$ gathered from the OC counter and the estimation result of the previous step in an expression. Dividing both numbers, the procedure yields the estimation for the current step. By iterating this calculation throughout all location steps of a path expression, the expression estimate is calculated.

This method introduces a strict dependency between the estimations of each step. For example, for estimating $//t/s/p$, we take three location steps $//t$, $/s$ and $/p$. The first one yields $occ(t) = 4$. For the second step, we probe the child spoke of ASPE(t) for an s and take its OC value, i.e., 3. Then, $occ(/s) = 4/3 = 1.5$. For the last step $/p$, we take the OC value of p in the child spoke of ASPE(s), i.e., 13. Thus, $occ(/p) = 13/1.5$ which is the estimated cardinality of the expression.

3 Empirical Evaluation

To assess the practical value of the EXsum extensions, we have systematically implemented and incorporated our ideas and competing summaries in our native XML database management system called XTC [9]. As competitor approaches, we have chosen XSeed [7] and LWES [2], whose parameter settings were adjusted as follows. For the XSeed kernel, we have set the search pruning parameter to 50 for the documents. For LWES, end-biased histograms were continuously applied to all levels of the summary structure.

We have used the following documents: *dblp* (330MB), *nasa* (25.8MB), *swissprot* (109.5MB) and *psd7003* (716MB). For each document, we have generated query workloads containing three basic query types: queries with *simplechild* and *descendant* path steps and those with predicates. In the case of EXsum and LWES, *parent* and *ancestor* queries are also evaluated, as they provide support for them. The workloads were processed on a computer equipped with an Intel Core 2 Duo processor running at 2.2 GHz and 3 GB of RAM memory, the Linux operating system, and Java 6. The XTC server process was running on the same machine.

To accurately compare the estimation quality of EXsum and its competitors in all our experiments, we have used an error metric called Normalized Root Mean Square Error (NRMSE) given by the formula $\sqrt{\sum_{i=1}^{n}(e_i - a_i)^2}/(\sum_{i=1}^{n}(a_i)/n)$, where n is the number of queries in the workload, e the estimated result size, and a the actual result size. NRMSE measures the average error per unit of the accurate result. Furthermore, we analyze estimation times and sizing, i.e., storage size and memory footprint needed for cardinality estimation of query expressions.

Table 1. Estimation times (in msec)

Simple child queries			
Doc.	EXsum	LWES	XSeed
dblp	2.85	3.18	13.21
nasa	3.55	3.30	11.60
swissprot	2.93	2.80	17.83
pds7003	3.86	3.15	3.28

Parent and ancestor queries		
Doc.	EXsum	LWES
dblp	4.39	7.00
nasa	4.42	4.50
swissprot	5.48	7.34
pds7003	4.00	3.34

Descendant queries			
Doc.	EXsum	LWES	XSeed
dblp	3.18	3.12	26.12
nasa	2.75	2.93	7.19
swissprot	2.95	3.20	20.00
pds7003	4.04	3.53	7.96

Queries with predicates		
Doc.	EXsum	XSeed
dblp	4.92	7.63
nasa	5.60	10.20
swissprot	11.80	24.84
pds7003	13.86	15.75

Timing Analysis. Estimation time refers to the time needed to deliver the cardinality estimations for a query addressing a given document, i.e., the time the estimation process needs to get the query expression, to access the summary (possibly more than once), and to report the estimate to the optimizer. Here, we report averages of the times needed for the queries in a workload.

Table 1 shows the estimation times classified by query types. As the timing differences among the EXsum's estimation procedures is negligible, we have reported in Table 1 just the worst results depicted in column *EXsum*. Obviously, EXsum delivers superior results for all document and query types; hence, its impact on the overall optimization process is very low. While LWES is comparable and XSeed is slightly slower for most queries, both of them consume prohibitive times for the estimation of queries with descendant axes. This makes us to infer that XSeed might provide unacceptable times if deeply structured documents come into play.

Table 2. Sizing Analysis

Storage (in KB)			
	EXsum		
Doc.	DPC	other	LWES	XSeed
dblp	7	6	2	7
nasa	9	9	2	7
swissprot	14	13	4	15
pds7003	7	7	2	6

Memory Footprint (in KB)			
Doc.	DPC	other	LWES	XSeed
# location steps = ceil(average depth)				
dblp	0.65	0.62	2	7
nasa	0.91	0.84	2	7
swissprot	0.68	0.65	4	15
pds7003	0.60	0.57	2	6
# location steps = maximum depth				
dblp	1.13	1.08	2	7
nasa	1.17	1.11	2	7
swissprot	0.82	0.78	4	15
pds7003	0.79	0.76	2	6

Sizing Analysis. The storage amount listed in Table 2 characterizes the size of a summary. LWES presents the most compact storage. XSeed is slightly more compact than EXsum for the compared documents.

Table 2 also compares the memory footprint for various estimation situations on all summaries/documents. We have computed the average memory size needed to estimate cardinalities for queries with two characteristics: queries whose number of location steps, whatever axes included, are equal to the document's average depth (rounded up to next integer value), and queries whose number of location steps is equal to the maximum document depth. These cases enable us to infer whether a summary needs to be entirely or only partially loaded into memory, i.e., whether or not the memory consumption of a summary is bounded by the number of location steps in a query during the estimation. Except for EXsum, all other methods require the entire structure in memory to perform cardinality estimations. EXsum, in contrast, only loads the referenced ASPE nodes and is, therefore, the summary with the lowest memory footprint and related disk IO. Thus, although the use of EXsum implies higher storage space consumption, the estimation process may compensate it by lower memory use and IO overhead.

Estimation Quality. In addition to the results presented, we also compared the estimation quality of our methods against LWES and XSeed. The comparison with the original interpolation method for EXsum has gained similar

Table 3. Child/Desc. Error(%)

Doc.	Prev.Step	DPC	LWES	XSeed
dblp	6.06	13.86	14.49	15.51
nasa	291.49	29.32	3.45	3.36
swissprot	202.55	0.00	12.10	10.01
pds7003	0.00	0.00	0.00	0.00

results; we have omitted it due to space restrictions. DPC is only applicable to child/descendant queries, whereas Prev.Step can be applied to all queries compared. XSeed does not support parent/ancestor queries and LWES does not support queries with predicates. Therefore, we cannot make a true Cartesian comparison.

We analyzed the accuracy of simple child and descendant queries in Table 3. We can see that EXsum itself delivers good estimations, except for cases where homonyms are scattered across the document (*nasa*, *swissprot*). In this case, Prev.Step cannot contribute with quality results. In general, however, DPC gains the best results for

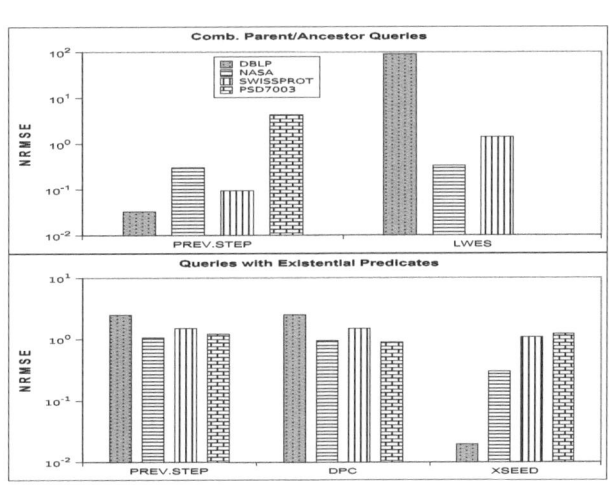

Fig. 5. Par./Anc. queries and predicates

the majority of cases and Prev.Step can provide high quality estimations on documents whose degree of structural variability is very low (*dblp* and *psd*7003). Furthermore, we have investigated the estimation quality for queries with parent and ancestor axes and queries with predicates (see Fig. 5). For the former, Prev.Step delivered high-quality estimations in most cases, comparable to or even better than LWES. Queries with predicates have obtained low estimation quality (an NRMSE reaching 100%). In this case, XSeed has a tendency to yield slightly better results in most of the cases and especially good ones for *dblp*.

4   Conclusion

In this paper, we have extended EXsum, an element-centered summary, to capture more statistical information on XML documents and, using this information, to support the estimation of a richer set of path expressions than possible with the base method. We have made a set of experiments to quantitatively evaluate our proposal against approaches published in the literature. Evaluating the new methods proposed for EXsum, DPC has delivered quality results for estimating structural path expressions, whereas the Prev.Step method has reported high errors concerning cardinality estimation for queries with child/descendant axis. For queries with parent/ancestor axes, Prev.Step has presented the best results. This empirical observation may lead us to infer that using only a single method to estimate all possible path expressions is not recommendable. More research is therefore needed to prove this finding.

References

1. Aboulnaga, A., Alameldeen, A.R., Naughton, J.F.: Estimating the selectivity of xml path expressions for internet scale applications. In: Proc. VLDB Conference, pp. 591–600 (2001)
2. Aguiar Moraes Filho, J.d., Härder, T.: Tailor-made xml synopses. In: Proc. BalticDB&IS Conference, pp. 25–36 (2008)
3. Freire, J., Haritsa, J.R., Ramanath, M., Roy, P., Siméon, J.: Statix: making xml count. In: SIGMOD Conference, pp. 181–191 (2002)
4. Lim, L., Wang, M., Padmanabhan, S., Vitter, J.S., Parr, R.: Xpathlearner: An on-line self-tuning markov histogram for xml path selectivity estimation. In: Proc. VLDB Conference, pp. 442–453 (2002)
5. Polyzotis, N., Garofalakis, M.N.: Xsketch synopses for xml data graphs. ACM Trans. Database Syst. 31(3), 1014–1063 (2006)
6. Wang, W., Jiang, H., Lu, H., Yu, J.X.: Bloom histogram: Path selectivity estimation for xml data with updates. In: Proc. VLDB Conference, pp. 240–251 (2004)
7. Zhang, N., Özsu, M.T., Aboulnaga, A., Ilyas, I.F.: XSeed: Accurate and fast cardinality estimation for xpath queries. In: Proc. ICDE Conference, p. 61 (2006)
8. Aguiar Moraes Filho, J.d., Härder, T.: EXsum—an xml summarization framework. In: Proc. IDEAS Symposium, pp. 139–148 (2008)
9. Haustein, M.P., Härder, T.: An efficient infrastructure for native transactional xml processing. Data Knowl. Eng. 61(3), 500–523 (2007)

Algorithm for Enumerating All Maximal Frequent Tree Patterns among Words in Tree-Structured Documents and Its Application

Tomoyuki Uchida and Kayo Kawamoto

Faculty of Information Sciences, Hiroshima City University, Hiroshima, Japan
{uchida,kayo}@hiroshima-cu.ac.jp

Abstract. In order to extract structural features among nodes, in which characteristic words appear, from tree-structured documents, we proposed a text mining algorithm for enumerating all frequent consecutive path patterns (CPPs) on a list W of words (PAKDD, 2004). First of all, in this paper, we extend a CPP to a tree pattern, which is called a *tree association pattern* (TAP), over a set W of words. A TAP is an ordered rooted tree t such that the root of t has no child or at least 2 children, all leaves of t are labeled with nonempty subsets of W, and all internal nodes, if exists, are labeled with strings. Next, we present text mining algorithms for enumerating all maximal frequent TAPs in tree-structured documents. Then, by reporting experimental results for Reuters news-wires, we evaluate our algorithms. Finally, as an application of CPPs, we present an algorithm for a wrapper based on CPP using XSLT transformation language and demonstrate simply the use of wrapper to translate one of Reuters news-wires to other XML document.

Keywords: text mining, tree structured pattern, maximal frequent tree pattern, wrapper.

1 Introduction

Many tree-structured documents contain large plain texts, have no absolute schema fixed in advance, and their structures may be irregular or incomplete. The formalization of representing knowledge is important for finding useful knowledge. We focus on the characteristics such as the usage of words and the structural relations among nodes, in which characteristic words appear, in tree-structured documents.

In [10], we introduced a consecutive path pattern (CPP, for short), which is a tree pattern representing structural features among leaves having characteristic words in tree-structured documents. First of all, by extending a CPP to a tree pattern, we define a *tree association pattern* (TAP, for short) over a set W of words as an ordered rooted tree t such that the root of t has no child or at least 2 children, all leaves of t are labeled with nonempty subsets of W, and all internal nodes, if exists, are labeled with strings. A TAP t is said to be *frequent* if t appears in a given set of tree-structured documents in a frequency of more than

D. Ślęzak et al. (Eds.): DTA 2009, CCIS 64, pp. 107–114, 2009.

a user-specified threshold. A frequent TAP t is *maximal* if there exists no frequent TAP which have t as a proper subpattern. In [10], the text mining algorithms for extracting all frequent CPPs from a given set of tree-structured documents were proposed. Moreover, Ishida et al. [6] presented online text mining algorithm for continuously extracting all frequent CPPs from an infinite sequence of tree structured documents. These algorithms were based on level-wise search strategy with respect to the length of a frequent CPP. Next, by modifying text mining algorithms in [10], we present an efficient text mining algorithm for enumerating all maximal frequent TAPs in a given set of tree-structured documents which appear in a given set of tree-structured documents in a frequency of more than a minimum support. The algorithm is based on level-wise search strategy with respect to the length of a frequent pattern. We use compact data structures such as FPTree presented by Han et al. [5] for managing all frequent TAPs. Then, in order to evaluate the performance of our algorithms, we apply our algorithm to Reuters news-wires [7], which contain 21,578 SGML documents and whose total size is 28MB. By reporting experimental results on our algorithm, we show that our algorithm have good performance for a set of a large number of tree-structured documents. Many researches focused on tags and their structured relations appearing in semistructured data [3,8,11]. But in this paper, we are more interested in words and their structural relations rather than tags in tree-structured documents.

Moreover, as an application of CPPs, we present an algorithm for a wrapper based on CPP using XSLT transformation language [2], which is an XML-based language used for transformation of XML documents into other XML documents. In order to show usefulness of our algorithms, we demonstrate briefly the use of wrapper to translate one of Reuters news-wires to other XML document.

2   Tree Association Pattern

For an integer $k \geq 2$, let $(w_1, w_2, \ldots, w_k)$ be a list of k words appearing in a given set of tree-structured documents such that words are sorted in ASCII-based lexicographical order. A *consecutive path pattern* on $(w_1, w_2, \ldots, w_k)$ (*CPP*, for short) has been introduced in [10] as a sequence $\langle t_1, t_2, \ldots, t_{k-1} \rangle$ such that for each i $(1 \leq i \leq k-1)$, t_i is a path between leaves labeled with $\{w_i\}$ and $\{w_{i+1}\}$, respectively. For example, the sequence $\alpha = \langle t_1, t_2, t_3 \rangle$ consisting of three paths t_1, t_2, t_3 given in Fig. 1 is a CPP on (COFFEE, SALVADOR, cocoa, week), which appears in document *xml_sample* in Fig. 1. For a set or a list S, the number of elements of S is denoted by $|S|$.

By extending a CPP to a tree pattern, for a nonempty subset W of $\mathcal{W}$ with $|W| \geq 2$, we call a node-labeled ordered rooted tree t a *tree association pattern* (*TAP*, for shot) over W if t satisfies either of the following conditions is satisfied. (1) t consists of only one node labeled with W. (2) The root of t has at least two child. Any leaf v and any internal node of t have a nonempty subset $W(v)$ of W and a string as labels, respectively, such that $W = \bigcup_{v \in V} W(v)$ for any two leaves u, v of t, where V is the set of all leaves of t. If $|W| = 2$, a TAP over W is called a

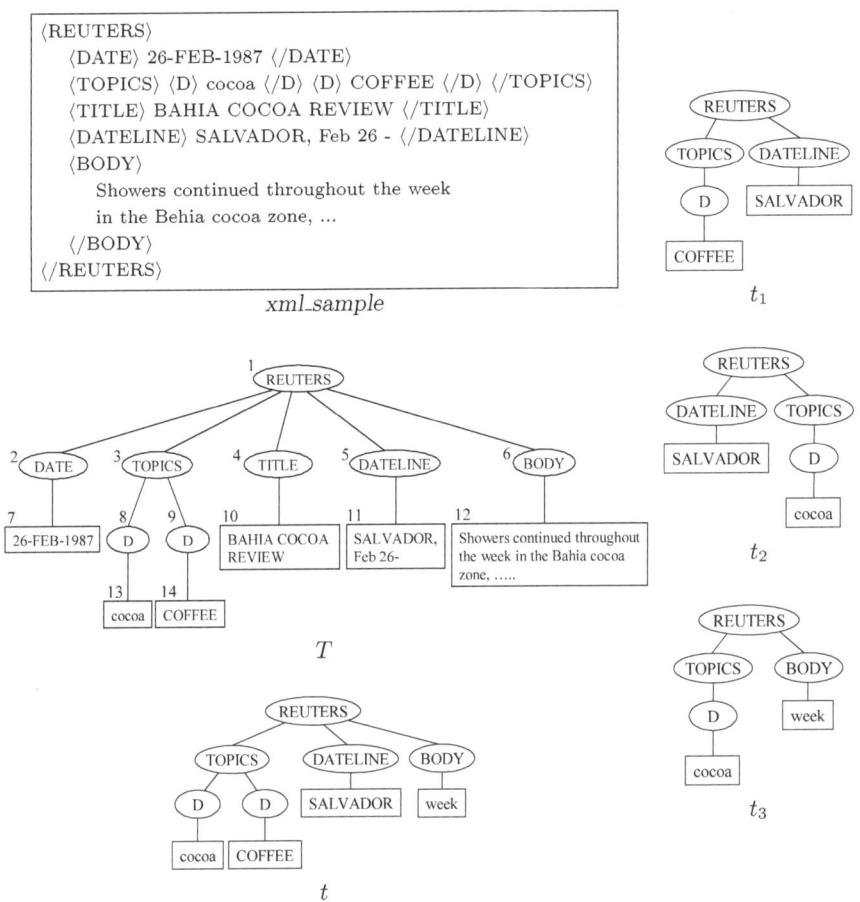

Fig. 1. Ordered tree T representing tree-structured data *xml_sample*, TAP t over {COFFEE, SALVADOR, cocoa, week}, and 2-trees t_1, t_2 and t_3 over {COFFEE, SALVADOR}, {SALVADOR, cocoa} and {cocoa, week}, respectively.

2-tree. As an example, we give a TAP t over {COFFEE, SALVADOR, cocoa, week} and 2-trees t_1, t_2 and t_3 over {COFFEE, SALVADOR}, {SALVADOR, cocoa} and {cocoa, week}, respectively.

A TAP t is said to *appear in* the ordered tree T_d, which represents a given tree structured document d, if there exists a subtree s of T_d such that t and s are isomorphic under the following conditions (1)-(3). This isomorphism from the node set of t to the node set of s is denoted by π. (1) v is a leaf of t if and only if $\pi(v)$ is a leaf of s. (2) For any internal node v of t, the labels of v and $\pi(v)$ are same. (3) For any leaf v of t, each word in the label of v appears in the label of $\pi(v)$ as a word. For example, TAP t and three 2-trees t_1, t_2, t_3 in Fig. 1 appear in the ordered-tree T representing the document *xml_sample* in Fig. 1. Let t' be a TAP over $W' \subseteq W$. The TAP t' is said to be a *subtree pattern* of

t if t' appears in t. Especially, a subtree pattern t' of t is said to be *proper* if $W' \subseteq W$ and $W - W' \neq \emptyset$.

3   Enumerating All Maximal Frequent TAPs from Tree-Structured Documents

Let D be a set of documents. For a TAP t, $Occ_D(t)$ denotes the number of documents in which t appear. For a set D of documents and a real number σ ($0 < \sigma \leq 1$), a TAP t is σ-*frequent* w.r.t. D if $Occ_D(t)/|D| \geq \sigma$. In general, a real number σ is given by a user and is called a *minimum support*. A σ-frequent TAP t is said to be *maximal* if there exists no σ-frequent TAP which has t as a proper subtree pattern. We formally define a data mining problem of enumerating all maximal frequent TAP as follows.

Maximal Frequent TAP Problem
Inst.   A set D of documents and a minimum support σ ($0 < \sigma \leq 1$).
Quest. Enumerate all maximal σ-frequent TAPs w.r.t. D.

In [10], we presented the level-wise search strategy for the problem of enumerating all σ-frequent CPPs from documents. By modifying algorithms in [10], we present an algorithm *FMF_TAP*, which based on a level-wise search strategy with respect to the length of a CPP, for solving Maximal Frequent TAP Problem in Fig. 2. In the same way as [10], in reading given documents, *FMF_TAP* constructs a *trie* (see [4]), which is a compact data structure, for efficiently storing and searching frequent words in given documents. In order to manage all frequent TAPs with respect to D, *FMF_TAP* uses an edge-labeled ordered rooted tree, called a *TAPTree*, which stores the information about a frequent TAP in the path from the root to node, and which is a compact data structure such as FPTree presented by Han et al. [5].

For a document d, $\mathcal{W}(d)$ denotes the set of all words appearing in d. For a set $D = \{d_1, d_2, \ldots, d_m\}$ of documents, $\mathcal{W}(D) = \bigcup_{1 \leq i \leq m} \mathcal{W}(d_i)$. For a document d and a word w, $\mathcal{AN}_d(w)$ denotes the set of all nodes of ordered tree representing d, whose value contains w as a word. For a set $D = \{d_1, d_2, \ldots, d_m\}$ of documents and a word w, let $\mathcal{ANS}_D(w) = \bigcup_{1 \leq i \leq m} \mathcal{AN}_{d_i}(w)$. At line 1, the set of all σ-frequent words with respect to D and $\mathcal{ANS}_D(w)$ for each σ-frequent word w are stored in the constructed data structure "trie". At line 2, *Make_2Tree* finds all frequent 2-trees by using *F_Word* constructed in line 1. At line 3, an edge-labeled ordered rooted tree T, whose depth is 1, is obtained by appending a new child v labeled with $(t, s, list)$ to the root of T, for each element $(t, s, list)$ in $\mathcal{F}$ constructed at line 2. At line 5, while there exist leaves in T which can be expanded, T is revised by appending new nodes to the leaves. *Make_2Tree* in line 2 and *Expand_TAPTree* in line 5 can be given by modifying algorithms *Make_2Tree* and *Expand_Pattern* in [10] to construct a TAPTree, respectively.

We implemented them in C++ on a PC running CentOS 5.2 with 3.0 GHz Xeon E5472 processor and 32GB of main memory. In the following experiments, as Stop Words, we chose symbols such as "-,+,...", numbers such as "0,1,2,...",

Algorithm **FMF_TAP**

Input: A set D of tree-structured documents, a minimum support σ $(0 < \sigma \leq 1)$ and a set Q of stop words.

Output: The set $\mathcal{F}$ of all maximal σ-frequent TAPs w.r.t. D.

begin

1. $F\_Word := \{(w, \mathcal{ANS}_D(w)) \mid w \in \mathcal{W}(D) - Q,\ w$ is σ-frequent w.r.t. $D\}$;
2. $\mathcal{F} := Make\_2Tree(F\_Word, \sigma)$;
3. Create the initial TAPTree T from $\mathcal{F}$;
4. Let H be the set of all leaves of T;
5. **while** $H \neq \emptyset$ **do** $(T, H) := Expand\_TAPTree(T, H, \mathcal{F}, \sigma)$;
6. Let $\mathcal{F}$ be the set of all maximal σ-frequent TAPs obtained from T;
7. **return** $\mathcal{F}$

end.

Fig. 2. Algorithm *FMF_TAP*

pronouns such as "it, this, . . .", articles "a, an, the", and auxiliary verbs "can, may, . . ." and so on. We apply these algorithms to Reuters-21578 text categorization collection in [7], which has 21,578 SGML documents and its size is about 28.0MB, in cases of each minimum support in {0.04, 0.06, 0.08, 0.10} and each number of documents in {5,000, 10,000, 15,000, 20,000, 21,578}. Fig. 3 (a) shows the running times of *FMF_CPP* in experiments. Each running time means a time to construct the TAPTree, which stores the information of all maximal frequent TAPs. In order to evaluate precisely performance of our algorithms, running time contains no reading time of input documents. Even if minimum support is 0.04, the running time is less than 2,300 sec. In every minimum supports, the running time is proportional regardless of the number of input documents. These show robustness of our algorithms. Fig. 3 (b) shows the numbers of all maximal frequent TAPs in each experiments. In case that minimum support is 0.04, our algorithm enumerate 15,018 maximal frequent TAPs within 2,211 sec., when 10,000 input documents are given. This shows the good performance of our algorithm. This is due to the data structure, TAPTree, which our algorithm adopts.

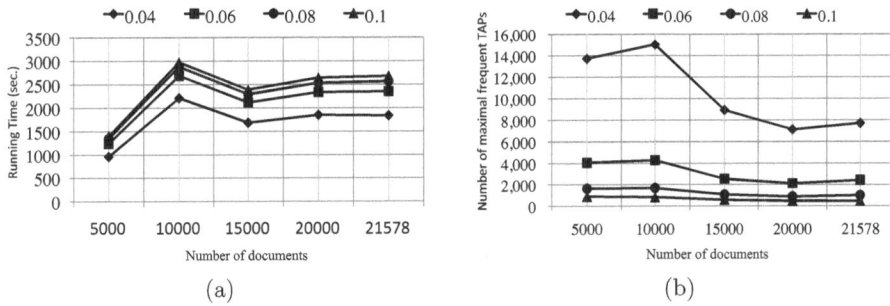

Fig. 3. (a) The running times of *FMF_TAP* and (b) the numbers of frequent maximal TAPs enumerated by *FMF_TAP*

4   Generating an XSLT Stylesheet Based on CPP

XSLT (eXtensible Style sheet Language Transformation) [2] is the W3C recommendation for an XML style sheet language and is an XML-based language used for the transformation of XML documents into other XML documents such as HTML documents and RTF documents. Document written by XSLT is called a *XSLT Stylesheet*. Software, which transforms a given XML document to other XML document by following a given XML Stylesheet, is called a *XSLT processor*.

Badica et al. [1] presented an approach for the efficient implementation of L-wrappers using XSLT transformation language, which is a standard language for processing XML documents. In a similar way as [1], we present an algorithm, *GEN_CPP_WRAPPER*, for a wrapper based on CPP using XSLT transformation language in Fig. 4. Let W be a set of words. For a tap t over W and CPP $\alpha = \langle t_1, t_2, \ldots, t_m \rangle$ over W such that α appears in t, π_t^α denotes a matching function from union of nodes in all 2-trees of α to set of nodes in t. The root r_i of t_i is called the *root* of α if $\pi_t^\alpha(r_i)$ is the root of t and for any j $(1 \leq j < i)$, $\pi_t^\alpha(r_j)$ is not the root of t. Number i is called a *key number* of α. In algorithm *GEN_CPP_WRAPPER*, procedures GEN-FIRST-TEMPLATE, GEN-TEMPLATE-WITH-VAR, GEN-TEMPLATE-NO-VAR and GEN-LAST-TEMPLATE, describe in [1] precisely, are as follows.

Algorithm **GEN_CPP_WRAPPER**(A CPP $\alpha = \langle t_1, t_2, \ldots, t_m \rangle$)

begin
1. Let r and ℓ be the root and the key number of α, respectively;
2. GEN-FIRST-TEMPLATE($[rn(t_\ell), r]$); // $[u, v]$ denotes path from u to parent of v;
3. $V = \emptyset$;
4. **for** $(i = \ell; i < m; i + +)$ **do begin**
5.   **if** the label of the parent of $rn(t_i)$ is a tag having attribute **then**
6.   **begin**
7. $var = $ GEN-VAR($rn(t_i)$);
8.   GEN-TEMPLATE-WITH-VAR($[ln(t_{i+1}, rt(t_{i+1})], [rn(t_{i+1}, rt(t_{i+1})], var, V)$;
          // $rt(t_{i+1})$ denotes the root of 2-tree t_{i+1}
9. $V = V \cup \{var\}$;
10.   **end**
11.   **else** GEN-TEMPLATE-NO-VAR($[ln(t_{i+1}, rt(t_{i+1})], [rn(t_{i+1}, rt(t_{i+1})], V)$;
12. **end**;
13. **for** $(i = \ell; i > 1; i - -)$ **do begin**
14.   **if** the label of the parent of $rn(t_{i-1})$ is a tag having attribute **then**
15.   **begin**
16. $var = $ GEN-VAR($rn(t_{i-1})$);
17.   GEN-TEMPLATE-WITH-VAR($[rn(t_{i-1}, rt(t_{i-1})], [ln(t_{i-1}, rt(t_{i-1})], var, V)$;
18. $V = V \cup \{var\}$;
19.   **end**
20.   **else** GEN-TEMPLATE-NO-VAR($[rn(t_{i-1}, rt(t_{i-1})], [ln(t_{i-1}, rt(t_{i-1})], V)$;
21. **end**;
22. GEN-LAST-TEMPLATE(V)
end.

Fig. 4. Algorithm *GEN_CPP_WRAPPER*

```
...
<xsl:stylesheet version="1.0" xmlns:xsl="http://www.w3.org/1999/XSL/Transform">
<xsl:template match="/">
 <result> <xsl:apply-templates select="/REUTERS/TOPICS/D" mode="selw1"/> </result>
</xsl:template>

<xsl:template match="*" mode="selw1">
 <xsl:variable name="var0"> <xsl:value-of select="."/> </xsl:variable>
 <xsl:apply-templates mode="selw2" select="parent::*/parent::*/DATELINE">
 <xsl:with-param name="var0" select="$var0"/>
 </xsl:apply-templates>
</xsl:template>
 ...
<xsl:template match="*" mode="display">
 <xsl:param name="var0"/>
 ...
 <xsl:value-of select="."/> </xsl:variable>
 <tuple>
  <d0><xsl:value-of select="$var0"/></d0>
  <d1><xsl:value-of select="$var1"/></d1>
  <d2><xsl:value-of select="$var2"/></d2>
  <d3><xsl:value-of select="$var3"/></d3>
 </tuple>
</xsl:template>
</xsl:stylesheet>
```

Fig. 5. Generated XSLT Stylesheet. We omit part of this Stylesheet, as it is too long to describe in full.

```
<result> <tuple>
  <d0>cocoa</d0> <d1>SALVADOR, Feb 26 - </d1> <d2>COCOA</d2>
  <d3>Showers continued throughout the week in the Behia cocoa zone,
  ...
 </d3>
</tuple> </result>
```

Fig. 6. Produced XML document. We omit part of this document, as it is too long to describe in full.

GEN-FIRST-TEMPLATE procedure generates the first template rule. GEN-TEMPLATE-WITH-VAR and GEN-TEMPLATE-NO-VAR procedures generate a template rule for paths, which are first and second arguments, respectively, depending if parent of start node of path has attribute or not. GEN-LAST-TEMPLATE procedure generates the last template rule. The constructing part of this rule fully instantiates the returned tree fragment, thus stopping the transformation process of tree representing input document.

Fig. 5 shows XSLT Stylesheet generated by applying $GEN\_CPP\_WRAPPER$ to CPP $\langle t_1, t_2, t_3 \rangle$ in Fig. 1. In Fig. 6, we show the XML document, which was produced by applying XSLT processor, Xalan, [9] to XML document xml_sample in 1 and XSLT Stylesheet in Fig. 5.

5   Concluding Remarks

In [10], we proposed a consecutive path pattern (CPP) as a sequence of paths to represent structural features among nodes in which given words appear in tree-structured documents. By extending a CPP to a tree pattern, we have defined a tree association pattern (TAP) over a set of words as a node-labeled ordered rooted tree that can represent structural features among nodes in which given words appear. Then, by modifying text mining algorithms in [6,10], we have presented text mining algorithms for enumerating all maximal frequent TAPs from a set of tree-structured documents. Moreover, as one of applications of CPPs we have presented a wrapper based on CPP using XSLT transformation language, which translate XML documents to other XML documents.

In order to design more effective search engine on Internet, we plan to develop our mining techniques based on CPPs and TAPs so that mining multimedia data such as Web contents with pictures, voice data and movie data.

References

1. Bădică, A., Bădică, C., Popescu, E.: Implementing logic wrappers using xslt stylesheets. In: Proc. ICCGI-2006, p. 31 (2006)
2. Clark, J.: XSL transformations (XSLT) version 1.0. W3C recommendation (1999), http://www.w3.org/TR/xslt
3. Fernandez, M., Suciu, D.: Optimizing regular path expressions using graph schemas. In: Proc. ICDE 1998, pp. 14–23 (1998)
4. Gonnet, G., Baeza-Yates, R.: Handbook of Algorithms and Data Structures. Addison-Wesley, Reading (1991)
5. Han, J., Kamber, M.: Data Mining: Concepts and Techniques. Morgan Kaufmann Publishers, San Francisco (2001)
6. Ishida, K., Uchida, T., Kawamoto, K.: Extracting structural features among words from document data streams. In: Sattar, A., Kang, B.-h. (eds.) AI 2006. LNCS (LNAI), vol. 4304, pp. 332–341. Springer, Heidelberg (2006)
7. Lewis, D.: Reuters-21578 text categorization test collection. UCI KDD Archive (1997), http://kdd.ics.uci.edu/databases/reuters21578/reuters21578.html
8. Miyahara, T., Shoudai, T., Uchida, T., Takahashi, K., Ueda, H.: Discovery of frequent tree structured patterns in semistructured web documents. In: Cheung, D., Williams, G.J., Li, Q. (eds.) PAKDD 2001. LNCS (LNAI), vol. 2035, pp. 47–52. Springer, Heidelberg (2001)
9. Apache Xalan Project. Xalan, http://xalan.apache.org/
10. Uchida, T., Mogawa, T., Nakamura, Y.: Finding frequent structural features among words in tree-structured documents. In: Dai, H., Srikant, R., Zhang, C. (eds.) PAKDD 2004. LNCS (LNAI), vol. 3056, pp. 351–360. Springer, Heidelberg (2004)
11. Wang, K., Liu, H.: Discovering structural association of semistructured data. IEEE Trans. Knowledge and Data Engineering 12, 353–371 (2000)

A Method for Learning Bayesian Networks by Using Immune Binary Particle Swarm Optimization

Xiao-Lin Li[1], Xiang-Dong He[2], and Chuan-Ming Chen[1]

[1] School of Business, Nanjing University,
Nanjing 210093, China
lixl@nju.edu.cn, ccming@nju.edu.cn
[2] Network and Information Center, Nanjing University,
Nanjing 210093, China
hexd@nju.edu.cn

Abstract. Bayesian network is a directed acyclic graph. Existing Bayesian network learning approaches based on search & scoring usually work with a heuristic search for finding the highest scoring structure. This paper describes a new data mining algorithm to learn Bayesian networks structures based on an immune binary particle swarm optimization (IBPSO) method and the Minimum Description Length (MDL) principle. IBPSO is proposed by combining the immune theory in biology with particle swarm optimization (PSO). It constructs an immune operator accomplished by two steps, vaccination and immune selection. The purpose of adding immune operator is to prevent and overcome premature convergence. Experiments show that IBPSO not only improves the quality of the solutions, but also reduces the time cost.

1 Introduction

The Bayesian belief network is a powerful knowledge representation and reasoning tool under conditions of uncertainty. Recently, learning the Bayesian network from a database has drawn noticeable attention of researchers in the field of artificial intelligence. To this end, researchers have developed many algorithms to induct a Bayesian network from a given database [1], [2], [3], [4], [5], [6].

Particle swarm optimization (PSO), rooting from simulation of swarm of bird, is a new branch of Evolution Algorithms based on swarm intelligence. The concept of PSO, which can be described with only several lines of codes, is more easily understood and realized than some other optimization algorithms. PSO has been successfully applied in many engineering projects.

This paper proposes a new data mining algorithm to learn Bayesian networks structures based on an improved swarm intelligence method and the Minimum Description Length (MDL) principle. An important characteristic of the algorithm is that, in order to prevent and overcome premature convergence, some concepts in immune systems are introduced into binary particle swarm optimization. Furthermore, the algorithm, like some previous work, does not need to impose restriction of having a complete variable ordering as input.

D. Ślęzak et al. (Eds.): DTA 2009, CCIS 64, pp. 115–121, 2009.
© Springer-Verlag Berlin Heidelberg 2009

This paper will begin with a brief introduction to Bayesian network and MDL principle. In section 3, the immune binary particle swarm optimization algorithm will be proposed. Then, in section 4 and 5, the performance of our algorithm will be demonstrated by conducting a series of experiments as well as a summary of the whole paper be made.

2   Bayesian Networks and MDL Metric

2.1   Bayesian Networks

A Bayesian network is a direct acyclic graph (DAG), nodes of which are labeled with variables and conditional probability tables of the node variable which is given its parents in the graph. The joint probability distribution (JPD) is then expressed in the following formula:

$$P(x_1, \cdots\cdots, x_n) = \prod_{k=1\cdots n} P(x_k \mid \pi(x_k)) \tag{1}$$

where $\pi(x_k)$ is the configuration of X_k's parent node set $\Pi(X_k)$.

Bayesian network Structure learning is a challenging problem. The main difficulty here is how to find a good dependency structure among the many possible ones, which are potentially and infinitely. For Bayesian networks, we know that the task of finding the highest scoring network is NP-hard [7].

2.2   The MDL Metric

The MDL metric [8] is derived from information theory and incorporates the MDL principle. With the composition of the description length for network structure and the description length for data, the MDL metric tries to balance between model accuracy and complexity. Using the metric, a better network would have a smaller score. Similar to other metrics, the MDL score for a Bayesian network, S, is *decomposable* and could be written as in equation 2. The MDL score of the network is simply the summation of the MDL score of $\Pi(X_k)$ of every node X_k in the network.

$$MDL(S) = \sum_k MDL(X_k, \Pi(X_k)) \tag{2}$$

According to the resolvability of the MDL metric, equation 2 can be written when we learn Bayesian networks form complete data as follows:

$$MDL(S) = N \sum_{k=1}^{N} \sum_{X_k, \Pi(X_k)} P(X_k, \Pi(X_k)) \log P(X_k, \Pi(X_k)) \tag{3}$$

$$- \sum_{k=1}^{N} \frac{\log N}{2} \parallel \Pi(X_k) \parallel (\parallel X_k \parallel -1)$$

Where N is database size, $\| X_k \|$ is the number of different values of X_k, and $\| \Pi(X_k) \|$ is the number of different parent value combinations of $\Pi(X_k)$.

3 Immune Binary Particle Swarm Optimization Method

PSO, originally developed by Kennedy and Elberhart [9], is a method for optimizing hard numerical functions on metaphor of social behavior of flocks of birds and schools of fish. It is an evolutionary computation technique based on swarm intelligence. A swarm consists of individuals, called particles, which change their positions over time. Each particle represents a potential solution to the problem. In a PSO system, particles fly around in a multi-dimensional search space. During its flight each particle adjusts its position according to its own experience and the experience of its neighbors, making use of the best position encountered by itself and its neighbors. The effect is that particles move towards the better solution areas, while still having the ability to search a wide area around the better solution areas. The performance of each particle is measured according to a predefined fitness function, which is related to the problem being solved and indicates how good a candidate solution is. The PSO has been found to be robust and fast in solving non-linear, non-differentiable, multi-modal problems. The mathematical abstract and executive steps of PSO are as follows.

Let the i th particle in a D -dimensional space be represented as $X_i = (x_{i1}, \ldots, x_{id}, \ldots, x_{iD})$. The best previous position (which possesses the best fitness value) of the i th particle is recorded and represented as $P_i = (p_{i1}, \ldots, p_{id}, \ldots, p_{iD})$, which is also called $pbest$. The index of the best $pbest$ among all the particles is represented by the symbol g. The location P_g is also called $gbest$. The velocity for the i th particle is represented as $V_i = (v_{i1}, \ldots, v_{id}, \ldots, v_{iD})$. The concept of the particle swarm optimization consists of, at each time step, changing the velocity and location of each particle towards its $pbest$ and $gbest$ locations according to Equations (4) and (5), respectively:

$$V_i(k+1) = \omega V_i(k) + c_1 r_1 (P_i - X_i(k))/\Delta t + c_2 r_2 (P_g - X_i(k))/\Delta t \qquad (4)$$

$$X_i(k+1) = X_i(k) + V_i(k+1)\Delta t \qquad (5)$$

where ω is the inertia coefficient which is a constant in interval [0, 1] and can be adjusted in the direction of linear decrease [10]; c_1 and c_2 are learning rates which are nonnegative constants; r_1 and r_2 are generated randomly in the interval [0, 1]; Δt is the time interval, and commonly be set as unit; $v_{id} \in [-v_{max}, v_{max}]$, and v_{max} is a designated maximum velocity. The termination criterion for iterations is determined according to whether the maximum generation or a designated value of the fitness is reached.

The method described above can be considered as the conventional particle swarm optimization, in which as time goes on, some particles become inactive quickly because they are similar to the *gbest* and lost their velocities. In the following generations, they will have less contribution for their very low global and local search capability and this problem will induce the emergence of the prematurity. Kennedy and Eberhart also developed the discrete binary version of the PSO. Then the particle changes its value by [11].

$$V_i(k+1) = \omega V_i(k) + c_1 r_1 (P_i - X_i(k))/\Delta t + c_2 r_2 (P_g - X_i(k))/\Delta t \qquad (6)$$

$$\text{if } \rho_i(k+1) < sig\left(v_i(k+1)\right) \text{ then } x_i(k+1) = 1 \ \Box \text{else } x_i(k+1) = 0 \qquad (7)$$

In this paper, an immune binary particle swarm optimization (IB-PSO) is proposed. In immune systems, for immature B cells the receptor editing is stimulated by B cell receptor and provides an important means of maintaining self-tolerance. The process of the receptor editing may diversify antibodies not only to jump local affinity optima, but also across the entire affinity landscape. In order to improve the performance of binary particle swarm optimization, we introduce the immune operator, which is accomplished by two steps, a vaccination and an immune selection. They are explained as follows.

Given an individual, a vaccination means modifying the genes on some bits in accordance with priori knowledge so as to gain higher fitness with greater probability [12]. A vaccine is abstracted from the prior knowledge of the pending problem, whose information amount and validity play an important role in the performance of the algorithm. But in most cases of dealing with some problems, it is difficult to abstract the characteristic information of them because we know little about the priori knowledge. On the other hand, the work of searching the local scheme used for the global solution makes the workload increase greatly and the efficiency decrease, so that the value of this work is lost. The proposed algorithm abstracts information from genes of the present optimal individual to make vaccines during the evolutionary process. Therefore, we can obtain vaccines easily.

The immune selection accomplish by the following two steps. The first one is the immune test. If the fitness is smaller than that of the parent, which means serious degeneration must have happened in the evolution, the parent will instead of the current individual and participate in the next generation to compete. The second one is the probability selection strategy based on antibody concentration. Define

$$D(x_i) = \frac{1}{\sum\limits_{j=1}^{N} |f_i - f_j|}, i = 1, 2, \ldots, N \qquad (8)$$

Where f_i is the fitness value of the i th particle, N is the number of the particles in the population, and $D(x_i)$ is the i th antibody concentration [13]. Define

$$P(x_i) = \frac{\sum\limits_{j=1}^{N}\left|f_i - f_j\right|}{\sum\limits_{i=1}^{N}\sum\limits_{j=1}^{N}\left|f_i - f_j\right|}, i = 1, 2, \ldots, N \qquad (9)$$

$P(x_i)$ is the probability selection formula based on antibody concentration.

The IBPSO method is summarized in table 1.

Table 1. The IBPSO method

---

1. Set to 0.
2. Create an initial population, *Pop(t)*. The initial population size is N.
3. If the current population contains the optimal individual, then the evolutionary stop; otherwise, continues.
4. Abstract vaccines according to the prior knowledge.
5. Create N offspring according to formula (6) and (7), then randomly create M new particles, and all new particles are stored in the intermediate population *Pop'(t)*..
6. Perform vaccination on *Pop'(t)* and obtain *Pop''(t)*.
7. Perform immune selection on *Pop''(t)* and obtain the next generation *Pop(t+1)*.
8. Go to 3.

---

4 Experiments

We conduct a number of experiments to evaluate the performance of the immune binary particle swarm optimization algorithm. The learning algorithms take the data set only as input. The data set is derived from ALARM network (http://www.norsys.com/netlib/alarm.htm).

Firstly, we generate 5,000 cases from this structure and learn a Bayesian network from the data set ten times. Then we select the best network structure as the final structure. We also compare our algorithm with binary particle swarm optimization and classical GA algorithm [6]. The MDL metric of the original network structures for the ALARM data sets of 5,000 cases is 81,219.74.

The population size N is 30 and the maximum number of generations is 5,000. We employ our learning algorithm to solve the ALARM problem. Some parameters in the experiment are taken as: $c_1 = 1.9$, $c_2 = 0.8$, $\omega = 0.5$, $\rho = 0.5$, $v_{max} = 8$. We also implemented a classical GA to learning the ALARM network. The one-point crossover and mutation operations of classical GA are used. The crossover probability p_c is 0.9 and the mutation probability p_m is 0.01. The MDL metric for immune binary particle swarm optimization algorithm, binary particle swarm optimization algorithm and the classical GA are delineated in Figure 1.

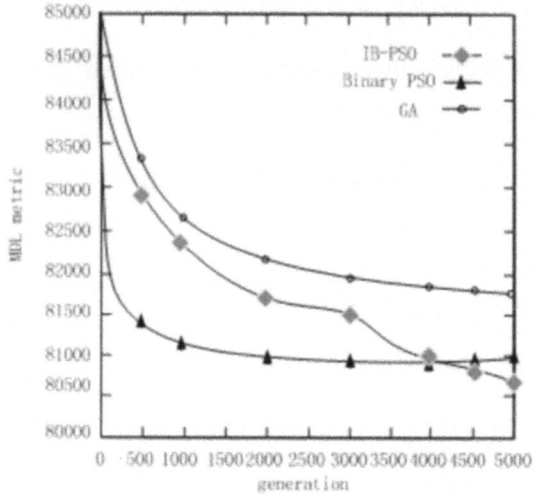

Fig. 1. The MDL metric for the ALARM network

From Figure 1, we see that the value of the average of the MDL metric for IB-PSO is 81223.3, the value of the average of the MDL metric for binary particle swarm optimization algorithm is 81268.3 and the value of the average of the MDL metric for the GA is 8,1789.4. We find immune binary particle swarm optimization algorithm evolves good Bayesian network structures at an average generation of 3991.6. Binary particle swarm optimization algorithm and GA obtain the solutions at average generation of 4083.7 and 4495.4, respectively. From Figure 1, we can also find that the proposed algorithm performs more poorly than Binary particle swarm optimization algorithm does at the early generations. But the performance of the proposed algorithm is better at the end of the generations. The reason of the phenomenon is that the proposed algorithm randomly creates M new particles in order to prevent and overcome the prematurity and use the best particle of the generation as the vaccination. Thus, we can conclude that immune binary particle swarm optimization algorithm finds better network structures at earlier generations than binary particle swarm optimization algorithm and the GA does.

5 Conclusions

Bayesian network is a directed acyclic graph. Existing Bayesian network learning approaches based on search & scoring usually work with a heuristic search for finding the highest scoring structure. This paper proposes an immune binary particle swarm optimization method by combining the immune system with PSO. In the IBPSO, the immune operator is designed for preventing and overcoming premature convergence. Then, the proposed method is used as a search algorithm to learn Bayesian networks from data. Experiments show that the proposed method is effective for learning Bayesian networks from data. Furthermore, studying the influence of the parameters for the evolutionary process on the performance of IBPSO is left for future work.

Acknowledgments

This work was supported by National Science Foundation of China (60803055), the China MOE Research Project of Humanities and Social Science (08JC630041), the China Postdoctoral Science Foundation (20080441031) and the Jiangsu Planned Projects for Post-doctoral Research Funds (0801038C).

References

1. Suzuki, J.: A construction of Bayesian networks from databases based on a MDL scheme. In: Proc of the 9th Conference of Uncertainty in Artificial Intelligence, pp. 266–273. Morgan Kaufmann, San Mateo (1993)
2. Xiang, Y., Wong, S.K.M.: Learning conditional independence relations from a probabilistic model, Department of Computer Science, University of Regina, CA, Tech Rep: CS-94-03 (1994)
3. Heckerman, D.: Learning Bayesian network: The combination of knowledge and statistic data. Machine Learning 20(2), 197–243 (1995)
4. Cheng, J., Greiner, R., Kelly, J.: Learning Bayesian networks from data: An efficient algorithm based on information theory. Artificial Intelligence 137(1-2), 43–90 (2002)
5. Lam, W., Bacchus, F.: Learning Bayesian belief networks: An algorithm based on the MDL principle. Computational Intelligence 10(4) (1994)
6. Larrañaga, P., Poza, M., Yurramendi, Y., Murga, R., Kuijpers, C.: Structure Learning of Bayesian Network by Genetic Algorithms: A Performance Analysis of Control Parameters. IEEE Trans. Pattern Analysis and Machine Intelligence 18(9), 912–926 (1996)
7. Chickering, D.M.: Learning Bayesian networks is NP-complete. In: Fisher, D., Lenz, H.J. (eds.) Learning from Data: Artificial Intelligence and Statistics V, pp. 121–130. Springer, Berlin (1996)
8. Lam, W., Bacchus, F.: Learning Bayesian belief networks: an algorithm based on the MDL principle. Computational Intelligence 10(4), 269–293 (1994)
9. Kennedy, J., Eberhart, R.C.: Particle swarm optimization. In: Proc. IEEE International Conference on Neural Networks, vol. 4, pp. 1942–1948 (1995)
10. Shi, Y., Eberhart, R.C.: A modified particle swarm optimizer. In: Proceedings of IEEE International Conference of Evolutionary Computation, Anchorage, Alaska, May 1998, pp. 69–73 (1998)
11. Kennedy, J., Eberhart, R.: A discrete binary version of the particle swarm optimization algorithm. In: Proceedings of the Conference on Systems, Man, and Cybernetics, pp. 4104–4109 (1997)
12. Jiao, L.C., Wang, L.: A novel genetic algorithm based on Immunity. IEEE Trans. on Systems, Man, and Cybernetics-Part A Systems and Humans 30(5), 552–561 (2000)
13. Lu, G., Tan, D.j.: Improvement on regulating definition of antibody density of immune algorithm. In: Proceeding of the 9th international conference on neural information processing, vol. 5, pp. 2669–2672 (2002)

A Semantics-Preserving Approach for Extracting OWL Ontologies from UML Class Diagrams

Zhuoming Xu, Yuyan Ni, Lili Lin, and Huajian Gu

College of Computer and Information Engineering, Hohai University,
Nanjing 210098, P.R. China
{zmxu,yyni_83,linlili,hjgu}@hhu.edu.cn

Abstract. Full implementation of the Semantic Web requires widespread availability of OWL ontologies. Manual ontology development using current OWL-aware editors remains a tedious and cumbersome task that can easily result in a knowledge acquisition bottleneck. Conceptual schemata such as UML class diagrams, on the other hand, capture abundant domain knowledge. Thus developing approaches and tools for extracting the available knowledge can facilitate the development of OWL ontologies and the construction of various Semantic Web applications. In this paper, a formal approach for extracting OWL DL ontologies from existing UML class diagrams is presented. The approach establishes a precise conceptual correspondence between the two models and relies on a semantics-preserving UML-to-OWL translation algorithm. Tool implementation and case study show that the proposed approach is effective and fully automatic ontology extraction from UML Class Diagrams is implementable.

Keywords: ontology development, schema translation, semantics preservation, UML class diagram, OWL ontology, Semantic Web.

1   Introduction

The success and proliferation of the Semantic Web [1] depends on quickly and cheaply constructing Web ontologies [2]. The importance of ontologies to the Semantic Web has prompted the establishment of the normative Web ontology language (OWL) [3] and the development of various OWL-aware ontology tools. Although the tools have matured over the years, manual ontology development using current ontology editors, such as Protégé, remains a tedious and cumbersome task that requires significant understanding of the new ontology language and can easily result in a knowledge acquisition bottleneck. So it is necessary to develop methods and tools that allow reducing the effort and adapting ontologies in a semi-automatic fashion using existing knowledge sources, being this the goal of ontology learning [4]. On the other hand, the OMG's Unified Modeling Language (UML) [5] has become a de facto industry standard for modeling applications in data and software engineering communities. It is a general-purpose visual modeling language that is designed to specify, visualize, construct and document the artifacts of a software system. Today,

D. Ślęzak et al. (Eds.): DTA 2009, CCIS 64, pp. 122–136, 2009.

many data sources have been modeled in UML class diagrams, in which abundant domain knowledge has been specified. Therefore, investigating the methods for translating the existing UML class diagrams into OWL ontologies is meaningful to the development of Web ontologies.

Another motivation of our work concerns the development of Semantic Web enabled applications. To build Semantic Web sites [6] or portals [7], existing Web data and documents must be "upgraded" to Semantic Web content which is semantically annotated with OWL ontologies. It is widely believed that the majority of current Web content is dynamic content powered by relational databases (RDB) [8]. Thus a critical requirement to realize the Semantic Web vision of "Web of Data" is the inclusion of the vast quantities of data stored in relational databases. The mapping of these vast quantities of data from RDB to the Resource Description Framework (RDF) (http://www.w3.org/RDF/) has been the focus of a large body of research work in diverse domains [9]. Investigations, such as [9], have shown that the incorporation of domain semantics through the use of application-specific domain ontology in mapping RDB to RDF is an important aspect to fully leverage the expressivity of RDF models – the standard data model of the Semantic Web. Therefore it is useful for us to develop methods and tools for extracting domain knowledge from database schemata (e.g., UML class diagrams) and mapping the knowledge to OWL ontologies.

Here we concentrate on UML class diagrams for the conceptual perspective. UML class diagrams allow for modeling, in a declarative way, the static structure of an application domain, in terms of concepts and relations between them. Several works [10][11][12][13] have investigated the mapping between UML class diagrams and Web ontologies. But most of the works have focused on extending UML and providing a visual method for modeling Web ontologies. As a result, users have to model UML class diagrams in an unfamiliar way and can not extract the existing knowledge from legacy UML class diagrams. A few researches [14][15] have addressed the problem of reusing knowledge in existing UML class diagrams to develop Web ontologies. However, to the best of our knowledge, these methods tend to extract limited knowledge from the UML model, without placing emphasis on the semantics-preservation of UML-to-OWL translation, and often use a non-normative ontology language or fail to give an easy-to-use tool.

In the paper, we investigate a more formal approach to automatic translation from a standard UML class diagram to an OWL DL (a sublanguage of OWL) ontology. The main idea of our approach is establishing a precise conceptual correspondence between the class diagram and the ontology and relying on a semantics-preserving translation algorithm. We also examine the implementablity of the proposed algorithm through the development of a prototype tool called UML2OWL, and test the execution efficiency of algorithmic routine via case study. The implementation and experiments show that the proposed approach is effective and fully automatic ontology extraction is implementable.

The remainder of the paper is organized as follows. Sect. 2 explicates the proposed UML-to-OWL translation approach. In Sect. 3 we present our prototype translation tool, UML2OWL, followed by the experimental results. We review related work in Sect. 4. Last section concludes our work.

2 Proposed Approach

To perform the extraction of OWL DL ontologies form UML class diagrams, we must formalize the two knowledge representation models, establish a precise conceptual correspondence between the two models and then design a semantics-preserving translation algorithm from UML class diagram to OWL DL ontology.

2.1 Formalization of UML Class Diagrams

The two models, UML class diagrams and OWL DL ontologies, share a set of core functionalities but despite this overlap, there are many features which can only be expressed in OWL, and others which can only be expressed in UML. Therefore, in our formalization of UML class diagrams, we restrict our attention to those aspects that constitute the core of a UML class diagram and are necessary for expressing ontological knowledge. The considered UML constructs and features are *class*, *association* or *association class* among classes, *attribute* of (association) class, association end (i.e., *role*), *multiplicity*, *generalization* between classes (excluding association classes) and *disjointness/covering constraints* enforced on generalization [5]. Other UML constructs and features are ignored because the target ontology language (i.e., OWL DL) does not have enough expressiveness or suitable constructs to capture their semantics (i.e., knowledge).

A UML class diagram can be formalized in term of the First Order Logic (FOL) or Description Logics (DLs) [16][17]. We adopt these ideas to specify our formalization of a UML class diagram. In the following, Definition 1 is an accessorial definition; Definition 2 gives the formal syntax and informal semantics of a UML class diagram.

Definition 1. For two finite sets X and Y, we call a function from a subset of X to Y an X-*labeled tuple over* Y. The labeled tuple Z that maps $x_i \in X$ to $y_i \in Y$, for $i = 1, 2, \ldots, k$ $(k \geq 1)$, is denoted $[x_1 : y_1, \ldots, x_k : y_k]$. We also write $Z[x_i]$ to denote y_i.

Definition 2. A *UML class diagram* is a tuple $U = (L_U, att_U, ass_U, mult_U, gene_U, dsjt_U, covr_U)$, where

- L_U is a finite *alphabet* partitioned into a set C_U of *class* (including *association class*) symbols, a set A_U of *attribute* symbols, a set S_U of *association* (including *association class*[1]) symbols, a set R_U of *role* (i.e., association end) symbols, and a set T_U of *data type* (primitive type) symbols; each $T \in T_U$ has an associated predefined *basic domain* $D(T)$.

---

[1] Note that in UML an association class is both an association and a class. Logically, the association and the association class are the same semantic entity, though they are graphically distinct in the class diagram; the semantics of an association class is a combination of the semantics of an ordinary association and of an ordinary class. Therefore, for an association attached with an association class, the association symbol in S_U and the association class symbol in C_U denote the same symbol (i.e., a single name) and represent the same underlying model element.

- att_U is a function that maps each class symbol in C_U to an A_U-labeled tuple over T_U. The function is used to model attributes of a class.

- ass_U is a function that maps each association symbol in S_U to an R_U-labeled tuple over C_U. We assume without loss of generality that: (i) each role is specific to exactly one association a class participates in; (ii) for each role $R \in R_U$, there is an association $S \in S_U$ and a class $C \in C_U$ such that $ass(S) = [...,R:C,...]$. The function associates a collection of roles to each association, determining implicitly also the *arity* of the association.

- $mult_U$ is a function from $C_U \times S_U \times R_U$ or $T_U \times C_U \times A_U$ to $N_0 \times (N_1 \cup \{\infty\})$ (where N_0 denotes non-negative integers, N_1 positive integers). The first component $minmult_U$ of $mult_U$ represents the lower bound and the second, $maxmult_U$, the upper bound. The function is used to specify *multiplicities*, i.e., constraints on the minimum and maximum number of times an instance of a class may participate in an association via some role, or on how many data values of a data type may fill an attribute of a class. If some multiplicity is stated as a single value in the class diagram, we consider both the lower and the upper bounds to be the same (i.e., equal to the value). If some multiplicity is not explicitly stated, then for roles, $minmult_U$ is assumed to be 0 and $maxmult_U$ be ∞; but for attributes, $minmult_U$ and $maxmult_U$ are all assumed to be 1.

- $gene_U \subseteq C_U \times C_U$ is an injective and acyclic binary relation over C_U that models a *generalization* relationship between a subclass (child class) and a superclass (parent class), meaning that each instance of the subclass is also an instance of the superclass. Several generalizations that have a common superclass can be grouped together to form a *generalization set*: $C_1 \ gene_U \ C$, $C_2 \ gene_U \ C$, ..., $C_m \ gene_U \ C$, where $C_1, C_2, ..., C_m, C \in C_U$, $m \geq 1$.

- $dsjt_U \subseteq C_U^1 \times C_U^2 \times ... \times C_U^m$, where $C_U^i = C_U, i = 1, 2, ..., m, m \geq 2$, is a *m*-ary relation over C_U that models a *disjointness constraint* enforced on a generalization set, meaning that the collection of subclasses in the generalization set have no instance in common; that is, their intersection is empty.

- $covr_U \subseteq C_U^1 \times C_U^2 \times ... \times C_U^m \times C_U$, where $C_U^i = C_U, i = 1, 2, ..., m, m \geq 2$, is a (*m*+1)-ary relation over C_U that models a *covering constraint* enforced on a generalization set, meaning that the collection of subclasses are covering for the superclass, i.e., every instance of the superclass is also an instance of at least one of its subclasses.

2.2 Definition of OWL DL Ontologies

OWL provides three increasingly expressive species (i.e., sublanguages) designed for use by specific communities of users and implementers [3][18]:

- *OWL Lite*, which supports users primarily needing a classification hierarchy and simple constraints.

- *OWL DL*, which supports users who want maximum expressiveness without losing computational completeness and decidability of reasoning systems.
- *OWL Full*, which is intended for users who want maximum expressiveness and the syntactic freedom of RDF without computational guarantees.

OWL DL has two types of syntactic form: *exchange syntax*, i.e., the RDF/XML syntax that is used to publish and share ontology data over the Web, and the frame-like style *abstract syntax* that is abstracted from the exchange syntax and thus facilitates access to and evaluation of the ontologies (being this the reason for describing our formal approach using the abstract syntax in the paper). Regardless of using which syntax form, the formal meaning of an OWL DL ontology is solely determined by the same underlying RDF graph of the ontology, which is interpreted by the *SHOIN(D)* Description Logic based model-theoretic semantics [18][19] for the language.

Typically, an OWL DL ontology consists of one optional ontology headers, which may include a set of ontology properties, plus any number of axioms including class axioms, property axioms, and individual axioms (i.e, facts about individuals). In the following, Definition 3 gives a concise definition of an OWL DL ontology which is suitable for capturing the knowledge extracted from a UML class diagram.

Definition 3. An *OWL DL ontology* is a tuple $O = (ID_O, Axiom_O)$, where

- ID_O is a finite OWL *identifier* set partitioned into the following subsets that are pairwise disjoint:
  - a subset CID_O of *class* identifiers including user-defined classes and two predefined classes **owl:Thing** and **owl:Nothing**,
  - a subset $DRID_O$ of *data range* identifiers; each data range identifier is a predefined XML Schema datatype reference such as **xsd:integer**,
  - a subset $OPID_O$ of *object property* identifiers; object properties link individuals to individuals, and
  - a subset $DPID_O$ of *datatype property* identifiers; datatype properties link individuals to data values.
- $Axiom_O$ is a finite OWL *axiom* set partitioned into a subset of *class axioms* and a subset of *property axioms*; each axiom is formed by applying OWL constructs (e.g., **Class** and **ObjectProperty**) to the identifiers or descriptions that are the basic building blocks of a class axiom and describe the class either by a class identifier or by specifying the extension of an unnamed anonymous class via the OWL 'Property Restriction' construct **restriction** and 'Boolean combination' construct **unionOf**.

Note that the complete syntax format for the identifiers in Definition 3 is a *URI reference* that consists of an absolute URI (or a prefix name) and a *fragment identifier*. Table 1 summarizes OWL DL constructs used here, their abstract syntax and model-theoretic semantics [19].

Table 1. The abstract/DL syntax and model-theoretic semantics of OWL DL constructs

OWL DL abstract syntax	DL syntax	Model-theoretic semantics
Descriptions (*C*)		
A (URI reference)	A	$A^I \subseteq \Delta^I$
owl:Thing	$\top$	owl:Thing$^I = \Delta^I$
owl:Nothing	$\bot$	owl:Nothing$^I = \varnothing$
restriction(R allValuesFrom(C))	$\forall R.C$	$(\forall R.C)^I =$ $\{x \mid \forall y. <x,y> \in R^I \to y \in C^I\}$
restriction(R minCardinality(n))	$\geq n\,R$	$(\geq n\,R)^I =$ $\{x \mid \#\{y. <x,y> \in R^I\} \geq n\}$
restriction(R maxCardinality(n))	$\leq n\,R$	$(\leq n\,R)^I =$ $\{x \mid \#\{y. <x,y> \in R^I\} \leq n\}$
restriction(U allValuesFrom(D))	$\forall U.D$	$(\forall U.D)^I =$ $\{x \mid \forall y. <x,y> \in U^I \to y \in D^I\}$
restriction(U minCardinality(n))	$\geq n\,U$	$(\geq n\,U)^I =$ $\{x \mid \#\{y. <x,y> \in U^I\} \geq n\}$
restriction(U maxCardinality(n))	$\leq n\,U$	$(\leq n\,U)^I =$ $\{x \mid \#\{y. <x,y> \in U^I\} \leq n\}$
unionOf(C_1 C_2 ...)	$C_1 \sqcup C_2$	$(C_1 \sqcup C_2)^I = C_1^I \cup C_2^I$
Data Ranges (*D*)		
D (URI reference)	D	$D^I \subseteq \Delta_D^I$
Object Properties (*R*)		
R (URI reference)	$R\,;\,R^-$	$R^I \subseteq \Delta^I \times \Delta^I\,;\,(R^-)^I = (R^I)^-$
Datatype Properties (*U*)		
U (URI reference)	U	$U^I \subseteq \Delta^I \times \Delta_D^I$
Class Axioms		
Class(A partial $C_1 ... C_n$)	$A \sqsubseteq C_1 \sqcap ... \sqcap C_n$	$A^I \subseteq C_1^I \cap ... \cap C_n^I$
EquivalentClasses($C_1 ... C_n$)	$C_1 = ... = C_n$	$C_1^I = ... = C_n^I$
DisjointClasses($C_1 ... C_n$)	$C_i \sqcap C_j = \bot\,, i \neq j$	$C_i^I \cap C_j^I = \varnothing\,, i \neq j$
Property Axioms		
ObjectProperty(R domain(C_1)...domain(C_n) range(C_1)...range(C_m)	$\geq 1\,R \sqsubseteq C_i$ $\top \sqsubseteq \forall R.C_j$	$R^I \subseteq C_i^I \times \Delta^I\,, i = 1,...,n$ $R^I \subseteq \Delta^I \times C_j^I\,, j = 1,...,m$
[inverseOf(R_1)])	$R = (^- R_1)$	$R^I = (R_1^I)^-$
DatatypeProperty(U domain(C_1)...domain(C_n) range(D_1)...range(D_m))	$\geq 1\,U \sqsubseteq C_i$ $\top \sqsubseteq \forall U.D_j$	$U^I \subseteq C_i^I \times \Delta_D^I\,, i = 1,...,n$ $U^I \subseteq \Delta^I \times D_j^I\,, j = 1,...,m$

2.3 Translation Algorithm

The rationale behind the automatic translation from a UML class diagram to an OWL DL ontology is the existence of semantic correspondences between the two models. These correspondences are briefly listed as follows:

- A UML package corresponds to an OWL ontology; the package name corresponds to the namespace for the ontology;
- A UML class corresponds to an OWL class (which can be called an *entity class*);
- A UML association (class) between two or more classes corresponds to an OWL class (which can be called a *relationship class*);
- A UML attribute of a class corresponds to an OWL datatype property whose domain is the OWL class corresponding to the UML (association) class; whose range is the predifined XML Schema datatype (e.g., **xsd:integer**) corresponding to the data type of this UML attribute;
- A UML role associated to an association and a class corresponds to an OWL object property between the two OWL classes that correspond to the UML association and class, respectively; the multiplicity an instance of the class may participate in the association via this UML role corresponds to the (minimum and maximum) cardinality of the OWL object property;
- A UML generalization set corresponds to a set of OWL **partial** (i.e., **subClassOf**) class axioms, each axiom corresponding to a generalization relationship. The disjointness and covering constraints enforced on the generalization set correspond to the **DisjointClasses** and **EquivalentClasses** axioms, respectively.

Note that in UML, the scope of an attribute is limited to the class on which it is defined, thus an attribute name is not globally unique within the diagram. Similarly, a UML association name can be duplicated in a given diagram, with each occurrence having a different semantics. However, in OWL all properties and classes are first class objects, meaning that they are defined globally and are uniquely identified by Uniform Resource Identifiers (URI) references. These incompatibilities should be tackled in the translation process by renaming and globalization of identifiers.

Based on the semantic correspondences, the formal approach for extracting an OWL DL ontology from a UML class diagram can be specified in the translation algorithm **U2OTrans** as follows. The formal proof for the semantics preservation of translation can be given. But here we leave it out for saving space.

Algorithm U2OTrans (U)

Input: A UML class diagram $U = (L_U, att_U, ass_U, mult_U, gene_U, dsjt_U, covr_U)$.

Output: The OWL DL ontology (in the abstract syntax) $O = (ID_O, Axiom_O) = \phi(U)$ that is defined by a *translation* ϕ such that $ID_O = \phi(L_U)$ and $Axiom_O = \phi(att_U, ass_U, mult_U, gene_U, dsjt_U, covr_U)$.

Steps:
1. The translation from UML symbols to OWL identifiers:
   1.1. for each class symbol $C \in C_U$, create a class identifier $\phi(C) \in CID_O$;
   1.2. for each association symbol $S \in S_U, S \notin C_U$, create a class identifier $\phi(S) \in CID_O$;

1.3. for each attribute symbol $A \in A_U$, create a datatype property identifier $\phi(A) \in DPID_O$;

1.4. for each data type symbol $T \in T_U$, create a data range identifier $\phi(T) \in DRID_O$, i.e., map it to a predefined XML Schema datatype reference;

1.5. for each role symbol $R \in R_U$, create two identifiers for a pair of mutually-inversed object properties: $\phi(R) \in OPID_O$ and $V = \mathbf{inv\_}\phi(R) \in OPID_O$, where ' $\mathbf{inv\_}$ ' is a terminal symbol part of the identifier.

2. The translation from UML elements to OWL axioms:

2.1. For each class $C \in C_U$ such that $att_U(C) = [A_1 : T_1, \ldots, A_h : T_h]$, $A_1, \ldots, A_h \in A_U$, $T_1, \ldots, T_h \in T_U$, $h \geq 1$,

    2.1.1. create a class axiom:

        Class($\phi(C)$ **partial**

            restriction($\phi(A_1)$ **allValuesFrom(** $\phi(T_1)$ **))) cardinality(1) ...**

            restriction($\phi(A_h)$ **allValuesFrom(** $\phi(T_h)$ **) cardinality(1))).**   (i)

    2.1.2. FOR $i = 1..h$ DO

        create a property axiom:

            DatatypeProperty($\phi(A_i)$ **domain(** $\phi(C)$ **) range(** $\phi(T_i)$ **);**   (ii)

        IF $(l \leftarrow minmult_U(T_i, C, A_i); l \neq 0)$ THEN replace the cardinality constraint in $\phi(A_i)$'s property restriction clause of axiom (i) with:

            minCardinality(l **);**   (iii)

        IF $(u \leftarrow maxmult_U(T_i, C, A_i); u \neq \infty)$ THEN replace the cardinality constraint in $\phi(A_i)$'s property restriction clause of axiom (i) with:

            maxCardinality(u **)).**   (iv)

2.2. For each association $S \in S_U$ such that $ass_U(S) = [R_1 : C_1, \ldots, R_n : C_n]$, $R_1, \ldots, R_n \in R_U$, $C_1, \ldots, C_n \in C_U$, $n \geq 2$,

    2.2.1. create a class axiom:

        Class($\phi(S)$ **partial**

            restriction($\phi(R_1)$ **allValuesFrom(** $\phi(C_1)$ **) cardinality(1)) ...**

            restriction($\phi(R_n)$ **allValuesFrom(** $\phi(C_n)$ **) cardinality(1))).**   (v)

    2.2.2. FOR $i = 1..n$ DO create a property axiom and a class axiom:

        ObjectProperty(V_i **domain(** $\phi(C_i)$ **) range(** $\phi(S)$ **) inverseOf(** $\phi(R_i)$ **));**   (vi)

        Class($\phi(C_i)$ **partial restriction(** V_i **allValuesFrom(** $\phi(S)$ **))).**   (vii)

2.3. For each role $R \in R_U$ such that $ass_U(S) = [\ldots, R : C, \ldots]$, $S \in S_U$, $C \in C_U$,

    2.3.1. create a property axiom:

        ObjectProperty($\phi(R)$ **domain(** $\phi(S)$ **) range(** $\phi(C)$ **);**   (viii)

    2.3.2. IF $(l \leftarrow minmult_U(C, S, R); l \neq 0)$ THEN add a cardinality constraint to V 's property restriction clause of axiom (vii):

            minCardinality(l **);**   (ix)

        IF $(u \leftarrow maxmult_U(C, S, R); u \neq \infty)$ THEN add a cardinality constraint to V 's property restriction clause of axiom (vii):

            maxCardinality(u **).**   (x)

2.4. For each generalization set C_1 $gene_U$ C, C_2 $gene_U$ C, ..., C_m $gene_U$ C, where $C_1, C_2, \ldots, C_m, C \in C_U$, $m \geq 1$,

2.4.1. FOR $i = 1..m$ DO create a class axiom:

    Class($\phi(C_i)$ **partial** $\phi(C)$ **);**   (xi)

2.4.2. IF $dsjt_U(C_1,...,C_m)$, THEN create a class axiom:

    DisjointClasses($\phi(C_1)...\phi(C_m)$ **);**   (xii)

    IF $covr_U(C_1,...,C_m,C)$, THEN create a class axiom:

    EquivalentClasses($\phi(C)$ **unionOf(** $\phi(C_1)...\phi(C_m)$ **)).**   (xiii)

Algorithm U2OTrans actually specifies a set of mapping rules following which a UML class diagram (Definition 2) can be translated into an OWL DL ontology (Definition 3). It is worth emphasizing that, although the syntax form of OWL axioms in the algorithm steps is the abstract syntax, it is possible for the programmer to design an algorithm implementation that generates an OWL ontology in the exchange syntax (i.e., the RDF/XML serialization) based on the mapping rules. This is because the correspondence between the two syntax forms is straightforward and has been established by W3C's OWL specification [18]. Let's take an example. The following is an instance of axiom (ii):

DatatypeProperty(staffNo domain(Staff) range(xsd:string))

This OWL abstract syntax's axiom corresponds to the following RDF/XML serialization:

```
<owl:DatatypeProperty rdf:ID="staffNo">
  <rdfs:domain rdf:resource="#Staff"/>
  <rdfs:range rdf:resource="http://www.w3.org/2001/XMLSchema#string"/>
</owl:DatatypeProperty>
...
<owl:Class rdf:ID="Staff"> ... </owl:Class>
```

Note that this serialization uses several predefined namespace prefixes (such as **owl:**) or the default namespace of the ontology to define ontological terms. The prefixes and default namespace are specified in the namespace declaration component of the ontology, like:

```
<!-- The default namespace of the current ontology: -->
xmlns = "http://www.uml2owl.org/university.owl#"
<!-- The base URI for the ontology document: -->
xml:base = " http://www.uml2owl.org/university.owl"
<!-- The namespace declaration for the "rdf:" prefix: -->
xmlns:rdf = "http://www.w3.org/1999/02/22-rdf-syntax-ns#"
<!-- The namespace declaration for the "rdfs:" prefix: -->
xmlns:rdfs = "http://www.w3.org/2000/01/rdf-schema#"
<!-- The namespace declaration for the "owl:" prefix: -->
xmlns:owl = "http://www.w3.org/2002/07/owl#"
```

Obviously, utilizing the specified correspondence between the abstract syntax and the exchange syntax [18] as shown in the example, our proposed algorithm is applicable to the generation of OWL ontologies in the RDF/XML syntax so that the ontology can then be directly published and shared over the Web.

3 Implementation and Experiments

As a proof-of-concept for the proposed approach (i.e., algorithm U2OTrans), we developed a prototype tool called UML2OWL based on J2SE 1.5.0 platform and conducted case study with the tool.

3.1 Prototype Tool

The UML2OWL tool can take an XML Metadata Interchange (XMI 2.0) coded UML class diagram file exported from a UML 2.0 modeling tool (such as No Magic, Inc.'s MagicDraw UML and Gentleware's Poseidon for UML) as input and produce the corresponding OWL DL ontology (in both the Abstract Syntax and the RDF/XML Syntax) as output. The architecture and process flow of UML2OWL is shown in Figure 1.

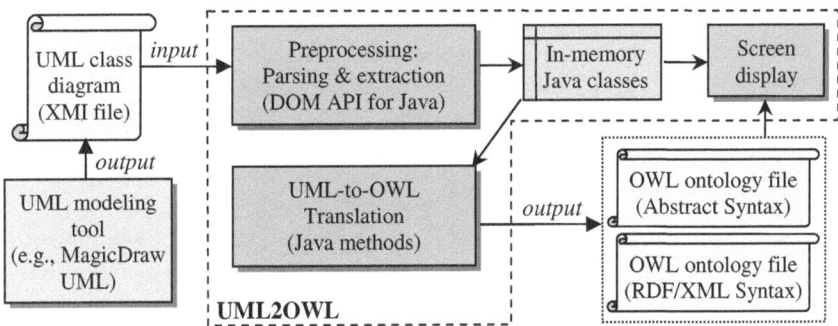

Fig. 1. Architecture and process flow of UML2OWL tool

In the implementation, the XMI file is parsed by the DOM API for Java. The extracted data from the XMI file are stored as in-memory data structure (Java classes) and simultaneously displayed in the tool screen. Then the in-memory data are used by some Java methods to perform the translation from the UML class diagram to the OWL DL ontology. The resulting ontology is saved as text files and displayed simultaneously in the tool screen.

3.2 Case Study

We carried out translation experiments of ten UML class diagrams using the implemented tool UML2OWL, with a PC (CPU P4/3.0GHz, RAM 512MB). The sizes of the diagrams range from 50 to 500. We use the total number of main elements including classes, attributes, associations, roles, and generalizations to measure the size N of a UML class diagram. All resulting OWL DL ontologies have passed the syntactic validation by the OWL Ontology Validator (http://phoebus.cs.man.ac.uk:9999/OWL/Validator). Case studies show that our approach is feasible and UML2OWL tool is efficient. Figure 2 shows the actual execution time of algorithmic routine in our case study, which indicates that the time complexity of translation is $O(N)$.

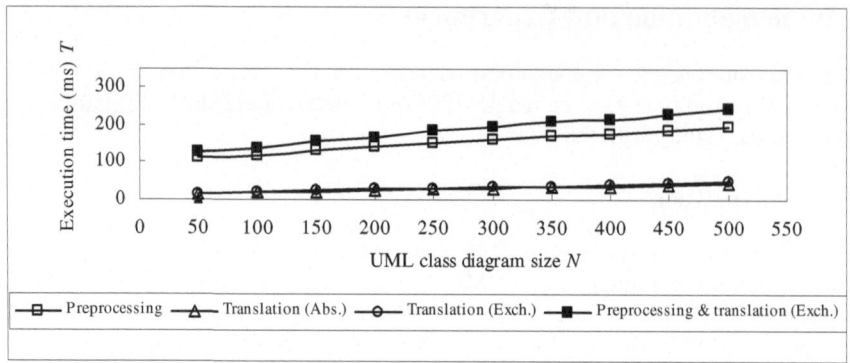

Fig. 2. Actual execution time of algorithmic routine on ten UML class diagrams

Saving space, Figure 3 shows a small-scale example UML class diagram, **Dept**, which models a university department and was created with No Magic, Inc.'s MagicDraw UML Community Edition v11.5 (http://www.magicdraw.com/).

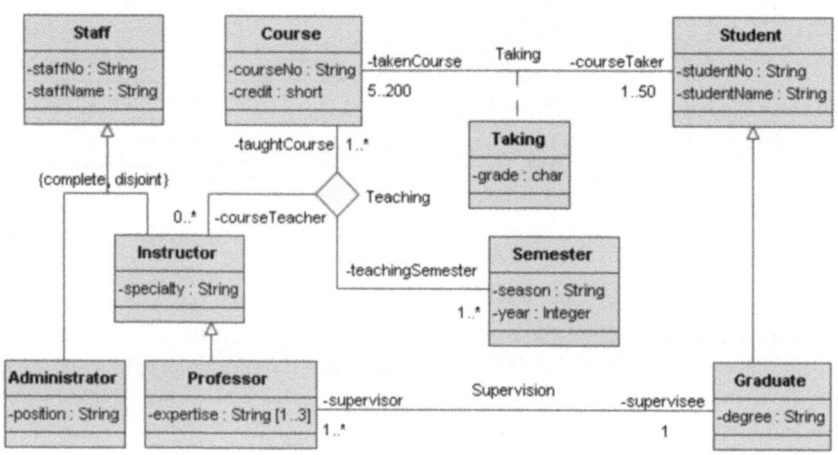

Fig. 3. UML class diagram **Dept** modeled with the MagicDraw UML tool, where '*' in the multiplicities denotes ' ∞ '

Figure 4 and Figure 5 are the screenshots of UML2OWL tool running the case in Figure 3. In Figure 4, the left tree is a visualization of the parsed UML class diagram, showing a UML class/association hierarchy with attributes/roles, where ⓒ denotes a class, Ⓢ an association, Ⓢ an association class, Ⓐ an attribute with its type, and Ⓡ a role with the association it participates in; the right text-area displays the resulting OWL DL ontology in the abstract syntax. In Figure 5, the right text-area displays the resulting OWL DL ontology in the RDF/XML Syntax.

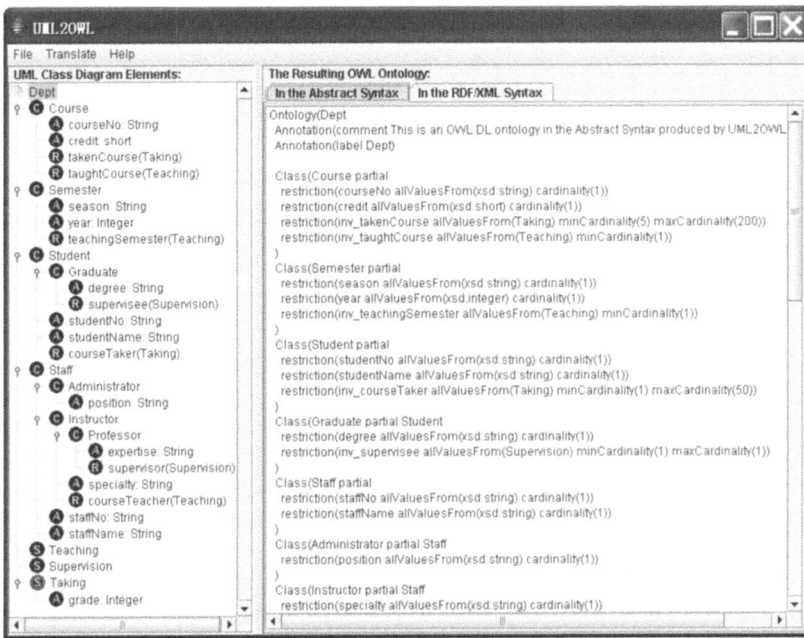

Fig. 4. Screenshot of UML2OWL, where the left displays the parsed UML class diagram **Dept** and the right shows the resulting OWL ontology in the abstract syntax

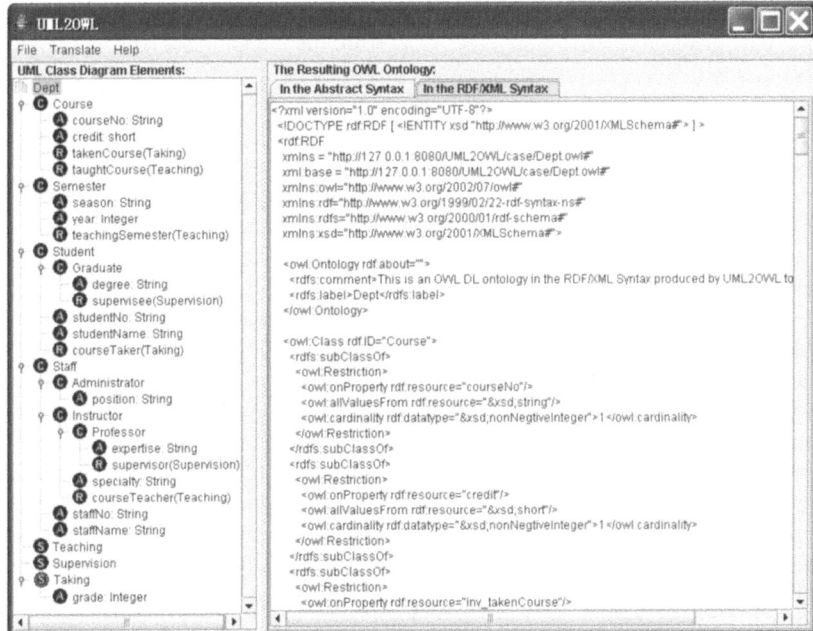

Fig. 5. Screenshot of UML2OWL, where the right shows the resulting OWL ontology in the RDF/XML syntax

4 Related Work

Several works have investigated the mapping between UML class diagrams and Web ontologies. They fall into two categories according to their focuses.

The first category focuses on *visual modeling*, i.e., *providing a UML-based visual environment for modeling Web ontology*. Cranefield [10] is the first who puts forward this idea. In his method, domain experts firstly create an ontology in a UML tool and then save it as an XMI-coded file. Finally an XSLT stylesheet translates the XMI-coded file into the corresponding RDF Schema (RDFS) ontology. Baclawski, *et al.* [11] proposed a method for extending UML to support ontology engineering in the Semantic Web. He summarized the similarities and differences between UML and DAML (an early non-normative Web ontology language) and proposed an extension of UML to resolve the differences. A mapping from the extended UML model to DAML ontologies was then defined in his work. The Components for Ontology Driven Information Push (CODIP) program (http://codip.projects.semwebcentral.org/) developed a tool called Duet that provides a UML-based environment for ontology development. The latest version of Duet can carry out the conversion between UML and OWL. As the standard UML can not completely express all OWL concepts, Djurić, *et al.* [12][13] introduced Ontology Definition Metamodel (ODM) in their research and defined an OWL-based UML profile for integrating software engineering into ontology development. Summing up these works, Cranefield's solution used RDFS as the language of the target ontology. However, this method has inherent drawbacks because RDFS does not have enough expressive power to capture the knowledge and constraints of UML class diagrams. All the other solutions used UML stereotype to extend the standard UML for bridging the gap between UML and Web ontology languages (DAML or OWL). As a result, the mapping sources will no longer be the standard UML model and each solution has its specific rules to create UML class diagrams – not the common and familiar way in which people model UML class diagrams today. Furthermore, ontology developers can not extract the existing domain knowledge from legacy UML class diagrams using these methods and tools.

The second category focuses on *reusing knowledge*, i.e., *extracting domain knowledge from existing UML class diagrams to facilitate the development of Web ontologies*. Our work belongs to this category. Along this line, the representative work is, among others, the research by Falkovych, *et al.* [14] and by Na, *et al.* [15]. Falkovych, *et al.* proposed the transformation from standard UML class diagrams to DAML+OIL (the predecessor of OWL) ontologies. This method converts UML classes into DAML+OIL classes, attributes of UML class into DAML+OIL datatype properties, UML binary-associations into DAML+OIL object properties (taking no account of UML associations' attributes). When converting UML associations, they first define four special object properties in DAML+OIL ("Binary", "Unidirectional", "Aggregation", and "Composition") according to the features of binary associations, and then map each UML association to a sub-property of one of these object properties. This solution has not considered *n*-ary associations. Besides, it can not use the ontology language to naturally model the conceptual entities and relationships which the UML descriptions are attempting to capture. Na, *et al.* [15] proposed a method for domain ontology building by extracting ontological elements from UML models designed previously. They compared the UML model elements with the OWL ones and derived transformation rules between the corresponding model elements. However, some important constraints in UML class diagrams, such as the disjointness constraints and covering constraints enforced on UML generalization sets, were ignored in the transformation. Furthermore, a formal algorithm and an easy-to-use tool have not been introduced in their work. In addition, the implementation of UML-to-OWL conversion relies on XSLT transformation which is quite cumbersome to use when one wants to express more complex mappings.

Compared with the existing solutions, our approach can capture richer knowledge of common UML constructs and constraints, use normative OWL language as the target ontology language, and develop a more formal, semantics-preserving algorithm and automated tool for extracting OWL ontologies from UML class diagrams.

5 Conclusions

Aiming to provide a common framework that allows data to be shared and reused across application boundaries, the Semantic Web has been proposed as the next generation of the current Web. Ontologies play an essential role in such an effort, providing a concise and systematic means for defining Web data semantics in a formal, machine-processable way. However, manual ontology development using current ontology editors remains a tedious and cumbersome task that can easily result in a knowledge acquisition bottleneck. Thus developing approaches and tools for extracting knowledge from existing knowledge models such as UML can facilitate the development of Web ontologies and the construction of various Semantic Web applications.

In this paper, a formal approach for extracting OWL DL ontologies from existing UML class diagrams is presented. The approach establishes a precise conceptual correspondence between the two models and relies on a semantics-preserving UML-to-OWL translation algorithm. Tool implementation and case study show that the proposed approach is effective and fully automatic ontology extraction from UML Class Diagrams is implementable. In the sense of semantic interoperability, the proposed approach and implemented tool can act as a gap-bridge between existing Web data sources and the Semantic Web.

Acknowledgements. This work was supported by (i) the "Qinglan Project" of Jiangsu Province of China, (ii) a grant (No. BK2008354) from the Natural Science Foundation of Jiangsu Province of China, and (iii) a grant (No. 2008135) from the "Six Talent Peaks Program" of Jiangsu Province of China. The views expressed in the paper represent those of the authors and not necessarily those of the funding bodies.

References

1. Berners-Lee, T., Hendler, J., Lassila, O.: The Semantic Web. Scientific American 284(5), 34–43 (2001)
2. Jacob, E.K.: Ontologies and the Semantic Web. Bulletin of the American Society for Information Science and Technology 29(4), 19–22 (2003)
3. Dean, M., Schreiber, G. (eds.): OWL Web Ontology Language Reference. W3C Recommendation (2004), http://www.w3.org/TR/owl-ref/

4. Maedche, A., Staab, S.: Ontology learning for the Semantic Web. IEEE Intelligent Systems 16(2), 72–79 (2001)
5. Object Management Group: Unified Modeling Language: Superstructure, Version 2.0 (2005), http://www.omg.org/docs/formal/05-07-04.pdf
6. Volz, R., Handschuh, S., Staab, S., et al.: Unveiling the hidden bride: deep annotation for mapping and migrating legacy data to the Semantic Web. Journal of Web Semantics 1(2), 187–206 (2004)
7. Stollberg, M., Lausen, H., Lara, R., et al.: Towards Semantic Web portals. In: Proc. of the WWW 2004 Workshop on Application Design, Development and Implementation Issues in the Semantic Web, CEUR Workshop Proceedings, vol. 105 (2004), http://CEUR-WS.org/Vol-105/
8. He, B., Patel, M., Zhang, Z., Chang, K.C.-C.: Accessing the Deep Web. Communications of the ACM 50(5), 94–101 (2007)
9. Sahoo, S.S., Halb, W., Hellmann, S., et al.: A Survey of Current Approaches for Mapping of Relational Databases to RDF. In: W3C RDB2RDF Incubator Group report (2009), http://www.w3.org/2005/Incubator/urw3/XGR-RDB2RDF-20090108/
10. Cranefield, S.: UML and the Semantic Web. In: Proc. of the first Semantic Web Working Symposium, pp. 113–130. Stanford University, California (2001)
11. Baclawski, K., Kokar, M.K., Kogut, P.A., et al.: Extending UML to support ontology engineering for the Semantic Web. In: Gogolla, M., Kobryn, C. (eds.) UML 2001. LNCS, vol. 2185, pp. 342–360. Springer, Heidelberg (2001)
12. Djuric, D., Gasevic, D., Devedzic, V.: Ontology modeling and MDA. Journal of Object Technology 4(1), 109–128 (2005)
13. Gasevic, D., Djuric, D., Devedzic, V.: MDA-based Automatic OWL Ontology Development. International Journal on Software Tools for Technology Transfer 9(2), 103–117 (2007)
14. Falkovych, K., Sabou, M., Stuckenschmidt, H.: UML for the Semantic Web: transformation-based approaches. In: Knowledge Transformation for the Semantic Web, Frontiers in Artificial Intelligence and Applications, vol. 95, pp. 92–106. IOS Press, Amsterdam (2003)
15. Na, H.-S., Choi, O.-H., Lim, J.-E.: A method for building domain ontologies based on the transformation of UML models. In: Proc. of 4th Int'l Conf. on Software Engineering Research, Management and Applications, pp. 332–338. IEEE Computer Society, Los Alamitos (2006)
16. Berardi, D., Calvanese, D., Giacomo, G.D.: Reasoning on UML class diagrams. Artificial Intelligence 168(1-2), 70–118 (2005)
17. Calvanese, D., Lenzerini, M., Nardi, D.: Unifying class-based representation formalisms. Journal of Artificial Intelligence Research 11, 199–240 (1999)
18. Patel-Schneider, P.F., Hayes, P., Horrocks, I. (eds.): OWL Web Ontology Language Semantics and Abstract Syntax. W3C Recommendation (2004), http://www.w3.org/TR/owl-absyn/
19. Horrocks, I., Patel-Schneider, P.F., van Harmelen, F.: From SHIQ and RDF to OWL: the making of a Web ontology language. Journal of Web Semantics 1(1), 7–26 (2003)

Data Warehousing and Business Intelligence: Benchmark Project for the Platform Selection

Cas Apanowicz

IT Horizon, Canada
cas.apanowicz@it-horizon.com

Abstract. The growth in importance of Data Warehousing (DW) and Business Intelligence (BI) has dramatically changed the IT landscape. The unprecedented rate of growth of information and raw data, coupled with ever accelerating escalation of competitiveness in business and government agencies, created never experienced before pressure on the IT executives to fulfill ever growing need of their internal clients for the prompt and accurate information. At the same time, new phenomena of open source have open opportunities to the medium and small companies of participating in the endeavor of BI. This newly revolutionized IT landscape seeks for a more systematic methodology of selecting platforms for DW/BI environment. The need for proper architecting of that environment with the need of balancing the performance level versus the Total Cost of Ownership (TCO) inspired us to propose a formalized model of benchmarking that would combine the transparency and rigorous formalism of the approach with practical guidelines of conducting highly objective process of measuring the most important performance parameters. Two main components will be the building blocks of the proposed method: A portable, customizable model of a database implemented in a new open source technology based on the columnar RDBMS offered by Infobright, as well as Evaluation Matrix equipped with parameters and weights that can be customized for any client. The paper discusses a practical implementation in a business case that was delivered for one of the Agencies of the Canadian Government.

Keywords: Data Warehousing, Business Intelligence, Benchmarking Practices, Product Selection, Evaluation Matrix, Preference Matrix, Priority Matrix. Infobright Enterprise Edition.

1 Introduction

One of the Canadian Government Agencies[1] has gone through extensive study of the Business Intelligence (BI) and Data Warehousing (DW) needs and preparedness assessment. There were several initiatives undertaken in researching DW/BI needs. Winter Corporation conducted the most prominent of them. Several recommendations were made for the upper management. One of them related to planning, designing and conducting a benchmark project along the following lines:

---

[1] For confidentiality reason, the name is removed from the paper.

D. Ślęzak et al. (Eds.): DTA 2009, CCIS 64, pp. 137–150, 2009.
© Springer-Verlag Berlin Heidelberg 2009

Develop an Enterprise BI vision

- Use BI pervasively to accomplish business objectives
- Better, more timely plans and decisions
- Enable more accurate predictions and forecasts
- React rapidly to events
- Yield better outcomes

Recruit Executive leadership

- Enterprise scope
- Business and technical scope

Involve business deeply

- Data and process knowledge
- They are the decision makers

Develop BI capabilities incrementally

- Individual projects contribute to enterprise effort
- Recognize and extend current capabilities
- Develop early successes and build on them

Integrate processes and data into Agency-wide BI perspective

- Understand process and data linkages to other initiatives and programs
- Understand what data will this project provide that can be used by others
- Understand what reports and analytics can be leveraged

The urgency of the above strategic principles created a need for development and adoption of technology vision and strategy. The fundamental cornerstone and critical factor in the success of this strategy is proper selection of a DW/BI platform. One of the options is to use the existing infrastructure. In order to properly assess the adequateness and suitability, as well as to identify any potential shortcomings of aforementioned platform, an appropriately designed evaluation process should be conducted. Such a process will become a necessary tool for comparative platform evaluation. In particular, it should identify any significant disparage between the user requirements and platform price/performance ratio. Such a process will also have to fulfill the following purposes:

- Provide a repeatable and objectively verifiable methodology supporting the platform selection
- Ensure adherence to the Industry Best Practices but be very user-centric
- Develop internal understanding of the Baseline that becomes a "Measuring Stick"
- Help to better understand DW/BI environment
- Identify software/hardware needs of the Agency through extensive study of DW/BI needs and preparedness.

1.1 Success Criteria

The purpose of Benchmarking Project is to present to The Agency a clear, simple, repeatable and objectively verifiable methodology, supporting the platform selection. Following this methodology will help to ensure that the process of designing and implementing the DW/BI platform will adhere to the best industry standards, and give the best performance in the most economical and cost-effective way to satisfy all stated agency DW/BI needs. It is also to map all business requirements - as articulated in the BI Readiness Assessment Strategy document prepared by Wintercorp Corporation and other body of work in this area – to the technology functionalities and needs. The third and final purpose of this document is to identify software/hardware needs for the benchmarking and estimate its cost. This document may also be used as an input into the detailed design phases of future projects to implement the proposed artifacts developed when conducting benchmarking strategy and process. The success criteria of the proposed strategy will be acceptance of this document. In particular, the agreement of the stakeholders on Evaluation Matrix, as it becomes the most measurable and objectively verifiable tool for conducting evaluation of any platform in the future. The separate issue is the success of the benchmarking process itself. Once the strategy is approved, it is of crucial importance to set the clear and understandable Key Success Factors of the process. They will, subsequently, become measuring indicators of the success of the entire process. The following are the aforementioned Key Success Factors:

- The process is well defined
- The timelines are agreed and followed
- All deliverables are completed
- User Requirements are extracted
- Requirements are mapped to Evaluation Matrix
- Testing environment (sand-box) is constructed:

    - Workload is identified
    - Query set is compiled
    - Data is extracted and populated
    - Testing processes are designed
    - Designed process is conducted

- Initial process is executed and Baseline is identified
- Gap analyzes are performed for each evaluated platform
- Evaluation Matrix is compiled

Meeting all above Key Success Factors, followed by the sign-off by delegated key stakeholders, will become a sufficient tool to properly measure and subsequently select an appropriate platform for the Data Warehouse/BI Environment.

2 Benchmarking Objectives

Hardware/Software and Cost Estimation: Identify software/hardware needs for the DW/BI platform and estimate its cost.

Provide Methodology: Provide repeatable and objectively verifiable methodology of conducting the benchmarking process. Once this process is defined and accepted by the agency, the process of platform selection - from the technological perspective – becomes more routine and well-documented process.

Platform selection Support: One of the most important purposes of creating the Benchmarking process is to simplify and support the process of platform selection. Although, by itself, it is not a selection procedure, the benchmarking process will address the two main aspects of the selection: technological suitability as well as cost-effectiveness. The additional elements – called in this document an *environmental* – have to be dealt with, outside of the scope of this document but the technological and cost-effectiveness support will be completely covered by benchmarking.

Adherence to the Industry Best Practices: The strategy of benchmarking described in this document will ensure that the selected Warehouse Strategy adheres to the Industry Best Practices but at the same time is very user-centric. This means that the technology being evaluated must – most of all – meet the user requirements and at the same time be the most flexible in meeting these needs.

Develop internal understanding of the Baseline: The properly conducted process will be instrumental in understanding the performance needs of the business user. At the same time the agency, as a whole, will learn more intimately the cost-effectiveness of the platform. This will come as a development of cost-performance *Baseline* – the basic performance requirements at the most optimal cost.

Better understanding of DW/BI environment: The processes of extraction of business requirements, designing the benchmark and creation of a - so-called - *sandbox*, and executing tests scenarios - will become a very valuable process of learning. During this time many assumptions and the understanding of agency needs will become solidified or modified. Many, before unforeseen elements, will surface or become more obvious. This entire learning process will be extremely valuable in fostering the best industry practices in DW and BI as well as refining the overall understanding of organizational needs. In particular it will bring better:

- Understand the queries
- Understand the volumetric
- Understand the network impact
- Understand architectural requirements

3 The Benchmark Approach

The adopted approach comprise of the number of steps and processes that will yield certain additional benefits outlined below.

Design and plan technology selection process: Employing the aforementioned methodology, the *Benchmark Strategy* document will be created to provide the layout, design and timelines of benchmarking process.

Identify Key Success Factors: The will produce a complete catalogue of factors, that will be used as a success criteria in the Benchmark process The client acceptance of these criteria, will indicate validity of the benchmarking process (e.g.: acceptance of Evaluation Matrix, acceptance of the methodology, etc.).

Create Evaluation Criteria: Create a *Priority Matrix* summarizing main area of interest based on which the final Evaluation Matrix will be created.

Create Evaluation Matrix: This matrix will have full list of attributes that will be measured during the benchmark process. Using the level of importance from Priority Matrix, each of these criteria will have appropriate weights assign to it base on Priority Matrix. This ensures that all criteria will affect the total score in the degree that is deemed appropriate by the customer. The combination of Priority Matrix and Evaluation Matrix, will allow every user to assign as much or as little importance to every evaluation parameter, providing the most flexible and most customizable measuring method.

Provide reusable artifacts for future development: Models, processes, codes, methodologies and practices that were will be developed and implemented during the Benchmark Process, can be reused directly or after some modification for the development of actual DW/BI platform.

Set up and execute query performance tests: Develop and implement a suite of queries, inserts and loads, which will be representative of the anticipated BI workload over the next five years. When delivered, the benchmark suite will be technology agnostic. It will therefore be capable of determining the technical capacity of any given platform to meet the upcoming BI based workload from the users.

Set up and conduct volumetric tests: Select and extract source data, modify and encrypt critical parts if necessary, increase original volume to the desired level to achieve a realistic production-like environment to better test load speed and capacity.

Select top leading candidates for evaluation: Create a list of alternative platforms that may potentially be a candidate for benchmarking.

4 The Benchmark Planning

The Industry Best Practices of benchmarking adhere to the benchmarking model based on the following steps:

1   **PLANNING**
Identify what is to be benchmarked
Identify comparative alternatives
Determine data collection method and collect data

2   **ANALYSIS**
Determine current performance levels
Project the future performance levels
Compile benchmark findings and gain acceptance

3   **INTEGRATION**
Establish functional goals
Develop action plans

4   **ACTION**
Implement specific actions and monitor progress
Recalibrate benchmarks

The proposed approach is very much in line with that spirit.

Requirements Extraction: The Agency has strong understanding of its needs and requirements in the area of Data Warehousing and Business Intelligence (DW/BI). This is mostly due to the fact that it has already undergone a lengthy process of gathering and documenting user requirement as they pertain to the Data Warehousing and Business Intelligence. There is a substantial repository of documents and research papers dealing with many aspects of DW/BI. Because of the above, the first step and main goal of this part of the benchmarking process, will be extracting all user requirements from the archrivals and repositories and compile a representative summary. Because of the nature and the form of collected User Requirements so fare as well as lack of complete documentation of the findings, , the main purpose of conducting user interviews and discussions with main stakeholders at this stage, will mostly be focused on confirmation and verification of material that will be extracted from all available documentation and personnel.

Requirements Analyses and Mapping: The next step in the process will be the analysis of all user requirements and mapping them to the technology functions. This part of the process can be presented in the following sequence of events:

Identify Data and Model: The first step while designing a benchmark is to examine the information framework of future DW/BI environment. Since the benchmarking process will have obvious limitation, the representative portion of this environment needs to be identified to serve as a benchmarking platform. Significant effort needs to be devoted to identifying the most optimal part of the production model that needs to be included in the testing/benchmarking. This comprises of two components: data selection and model definition. This model should be as close to production to facilitated user originated queries that actually take place in everyday operation and at the same time complex enough to properly measure required performance. Special care should be given to the selection of data sources. This should reflect future needs of the Data Warehouse and be fully representative of the volume, heterogeneity and the complexity of access in real production environment.

Identify ETL Process: Once the benchmark model and data sources are identified, the process of mapping the source to the target model needs to be conducted. The result of this process, should a fairly representative collection of ETL processes and tasks. Special care should be given to create ETL process that will handle high volume, complexity of the sources and be as close as possible to the final specification of DW/BI as stated in the user requirements documents. This way a significant portion of the ETL development can be re-used in the actual constriction of the DW. The following are the main goals of this task:

1. Emulation of the future ETL workload
2. Examine performance in the most demanding scenarios
3. Examine integration with the DW platform
4. Achieve the highest possible level of reusability

Identify queries: Similarly, all queries that will be collected in this process needs to reflect the current production needs as well as the requirements articulated during requirements gathering and system analyses. The main goals of query selection are:

1. Collect the most representative queries related to the OLTP system
2. Collect queries that are closest representation of the current/future BI needs
3. Collect the most complex queries
4. Collect the most frequently used (or expected to most frequently used)
5. Collect query involving the most volume

The reason for this selection is to examine the system behavior and needs under the most demanding circumstances.

Identify initial Functional Priorities: During this phase, all the documentation will be reviewed in order to extraction all functional needs and requirements. As a result of this process a list of queries, ETL processes and transformations, database operations, loads, restores and other similar functions will be prepared. All required batch windows, transactions and real-time data feeds will also be documented. This process will also verify the priorities assigned to them in Evaluation Matrix. The output of this exercise will take form of functional matrix compiling total list of all mentioned artifacts as presented in Evaluation Matrix. Let us also note that, at this point, the matrix is only a starting point, and as the process of *User Requirements Extraction* and its verification progress, more items can be added to the above list or the priorities may change.

Identify initial Performance Priorities: After completing the inventory of functions, all the functionalities collected during that process would need to have performance Baseline assigned. Every query, load process and similar, will have to have the expected performance assigned based on user needs. This phase of user requirements study will be aimed at this type of estimation: mapping user requirement to functionality performance metrics. The performance requirements can be illustrated by the following simple example. Consider the following scenario:

1. **A** - is a business Activity that needs to be done
2. **Q** – is a query supporting the action A
3. **T1** – is the time by which the activity A needs to be completed
4. **T2** – Time required for analyzing the results of query Q
5. **T3** – Time that it takes to perform activity A
6. **T4** – Execution time of query Q

Then, there may a business requirement that:

$$T4 < T1 - (T2+T3).$$

Identify Volumetric: The similar process will be conducted with respect to the volume. Based on the gathered earlier and verified during this process, all the volumetric needs will be documented. Item such as current volume of data, estimated or forecasted growth in 6 month, 1, 2, 3 and 5 years increments; Number of concurrent queries, number of concurrent users and the expected growth of these numbers need to be documented and then used for performance test cases.

Identify Scalability Needs: Similarly to the performance and volumetric, the scalability needs have to be documented and incorporated to the benchmarking process. The following effect on the platform needs to be addressed:

- Increased Data Volume
- Increased query complexities
- Increased query number

It is important to remember that almost all scalability issues can be addressed by compiling an enormous amount of hardware and software. But the real measurement of the scalability lays in the cost-effectiveness of the proposed solution to the expected growth. If two alternative platforms required additional cost to address expected growth: amount $X and amount $Y, then it obvious that the solution required less additional cost is more scalable; even though the both can cope with the increase at the satisfactory level.

As described before Priority Matrix and Evaluation Matrix will be the base for the platform evaluation. Priority Matrix will be set to reflect requirements priority at the highest level. It will serve as guidelines for compiling and tabulating the final results in Evaluation Matrix. This is the high level summary of user preferences. It lists the number of broad criteria that will be further refined in Evaluation Matrix. Each item from Priority Matrix will become a group in Evaluation Matrix and will be expanded further in more details. Relative importance will be used to weight the scoring in Evaluation Matrix. The Total value of all relative weights in Priority Matrix is 100.

Table 1. Priority Matrix used for weight assignment in the Architecture Selection Matrix

Area of Interest	Description	Relative Importance
Completeness of the solution	*The solution should be comprehensive and adhere to the Industry Best Practice. Provide right platform and format for future development end evolution. Following the industry trend.*	**5.0**
Performance	*Being able to accommodate expected growth of volume, users, ways of usage and type of collected sources. Scale: 0 - Exponential, 1 - Above Linear, 3 - Linear, 5 - Sub-linear*	**25.0**
Scalability	*The solution should be comprehensive and adhere to the Industry Best Practice. Provide right platform and format for future development end evolution. Following the industry trend.*	**25.0**
Time to Implementation	*How long does it take from initiation, through planning, designing, building to production implementation and running? How quickly can it be used?*	**20.0**
Information/ Knowledge Completeness	*Does the solution give full access to the full and comprehensive information and knowledge? Does the solution fully utilize the data, information and knowledge? Does the solution fully satisfy user requirements?*	**5.0**
Vendor Reliability and Dependency	*Is the vendor reputable? How well is vendor accountable? Is the vendor reliable? (How well and quickly responds to the requests and calls?) Does vendor care?.*	**10.0**
Support	*Availability of skills and expertise on the market (how difficult and expensive is to higher specialists).*	**6.0**
Platform Housekeeping	*Item such as backup recovery, archival, compression, etc*	**4.0**
TOTAL		**100**

Evaluation Matrix contains actual scoring of all relevant attributes of the evaluated platform. Each group has its own score weighted by the Relative Importance weight from Priority Matrix (e.g.: if total score of items from Support group is 100 – it will be multiply by Relative Importance = 0.05, so the total score will be 5).

Table 2. Comparison Matrix – evaluation grid based on Agency's requirements

Measured Attribute	Description	Score
Essential Attributes	*Any given platform must pass the minimum requirements in this category in order to be afforded any points in the non-essential category.*	

Table 2. (*continued*)

Measured Attribute	Description	Score
Query Performance	**Essential**	**1000**
Query Complexities	**Essential**	**500**
Parallel Queries (Multi-user)	**Essential**	**95**
Parallel Execution of a single query	**Desirable**	**40**
Query Prioritization	**Desirable**	**25**
ETL support	**Essential**	**300**
Load	**Essential**	**2000**
Real- or near-real Time Feed	**Essential**	**300**
Diversity of sources data	**Essential**	**300**
Volume Increase	**Essential**	**300**
Workload Increase	**Essential**	**300**
Provides data compression / decompression	**Essential**	**500**
Query Performance	**Essential**	**1000**
Query Complexities	**Essential**	**500**
Parallel Queries (Multi-user)	**Essential**	**95**
Parallel Execution of a single query	**Desirable**	**40**
Non-Essential Attributes	*These attributes will be afforded only to a platform that meets the required criteria specified in the Essential Attributes. If a platform performance on the Essential category is less than satisfactory, the values in this category will be set to zero.*	
Relative Value of the one month of implementation	**Essential**	**100**
Estimated # of Months before RFU	**Essential**	**8**
*The final Score for this attribute is calculated as follow: First we assume - based on experience - that estimated number of Months before User can query DW is 24. Then the final score for this attribute is = (24 - Estimated number of month before RFU) * score weight of 1 moth * Importance weight from the Importance Matrix * Z where Z = 3 if attribute is Essential, 2 if attribute is Desirable and 1 if nice to have.*		
Support for BI of OLTP System	**Essential**	**1000**
Data Sources		**290**
Supports source extract from multiple legacy and warehouse platforms, incl. Z/OS, DB2, VSAM, other	**Essential**	**50**

Table 2. (*continued*)

Measured Attribute	Description	Score
Can utilize source log and/or journals to determine net-change (specifically appropriate in CDC or CFT being supported by RDBMS platform)	**Nice to have**	30
Performance of copy and/or replication	**Nice to have**	90
Comprehensive support of transformation functions: validate, translate, integrate, check integrity, calculate, derive, aggregate, allocate and other	**Essential**	85
Provides loading of target via replace, append, update functions and bulk load on target RDBMS	**Essential**	85
Job Control and Scheduling		225
Supports job scheduling (time / event-based)	**Desirable**	75
Supports monitoring, reporting, recovery from unsuccessful completion	**Desirable**	75
Supports archive / restore	**Nice to have**	25
Supports security via access and encryption checks	**Desirable**	25
Provides time synchronization of updates	**Nice to have**	25
Metadata and Standards		195
Driven by open repository	**Nice to have**	50
Supports metadata exchange with other key products	**Nice to have**	50
Supports messaging and communication standards, including: XML, ODBC, etc.	**Nice to have**	70
Supports transport levels, including TCP/IP, FTP	**Nice to have**	25
Backup/Recovery	**Desirable**	200
24/7 Availability	**Desirable**	100
Professional skills Availability	**Essential**	100
Software Interoperability	**Essential**	100
Availability of Publications	**Essential**	100
Platform Stability	**Essential**	100
Tech support	**Essential**	80
Documentation	**Desirable**	80
Training availability / quality	**Desirable**	70

Table 2. (*continued*)

Measured Attribute	Description	Score
Consulting availability / quality	**Desirable**	**50**
Technical support responsiveness	**Essential**	**100**
Reliability	**Essential**	**60**
Accountability	**Essential**	**100**
Support for BI of OLTP System	**Essential**	**1000**
Data Sources		**290**
Supports source extract from multiple legacy and warehouse platforms, incl. Z/OS, DB2, VSAM, other	**Essential**	**50**
Can utilize source log and/or journals to determine net-change (specifically appropriate in CDC or CFT being supported by RDBMS platform)	**Nice to have**	**30**
Performance of copy and/or replication	**Nice to have**	**90**
Comprehensive support of transformation functions: validate, translate, integrate, check integrity, calculate, derive, aggregate, allocate and other	**Essential**	**85**
Provides loading of target via replace, append, update functions and bulk load on target RDBMS	**Essential**	**85**
TOTAL		**10943**

LEGEND

Evaluation criteria = Outline requirements * Requirement weighting
Requirement weighting => Essential = x3, Desirable = x2, Nice to have = x1

Additionally, the Requirements Weights will be multiplied by factors based on the degree of satisfying said requirements as follow:

0 = does not meet requirements
1 = partially meets requirements
2 = meets requirements
3 = exceeds requirements

Points = (weighting * importance from Importance Matrix) * evaluation score

Evaluation score will be based upon reviewing and analyzing platforms where the base-line scores will be considered a 100% in all respective categories. All other benchmarking platforms will be measured as a percentage of the above. The detailed method of measuring each parameter will be finalized during the benchmark process.

The Baseline in the presented case was implemented in the Infobright DBMS server – the only Open Source DW appliance and one that yielded spectacular Baseline performance. Additional reason for using the Infobright server was its ability to host vast amount of data in relatively small amount of SAN space (Infobright compression ratio of 10:1 allowed to have 2TB of data in 200GB disk space).

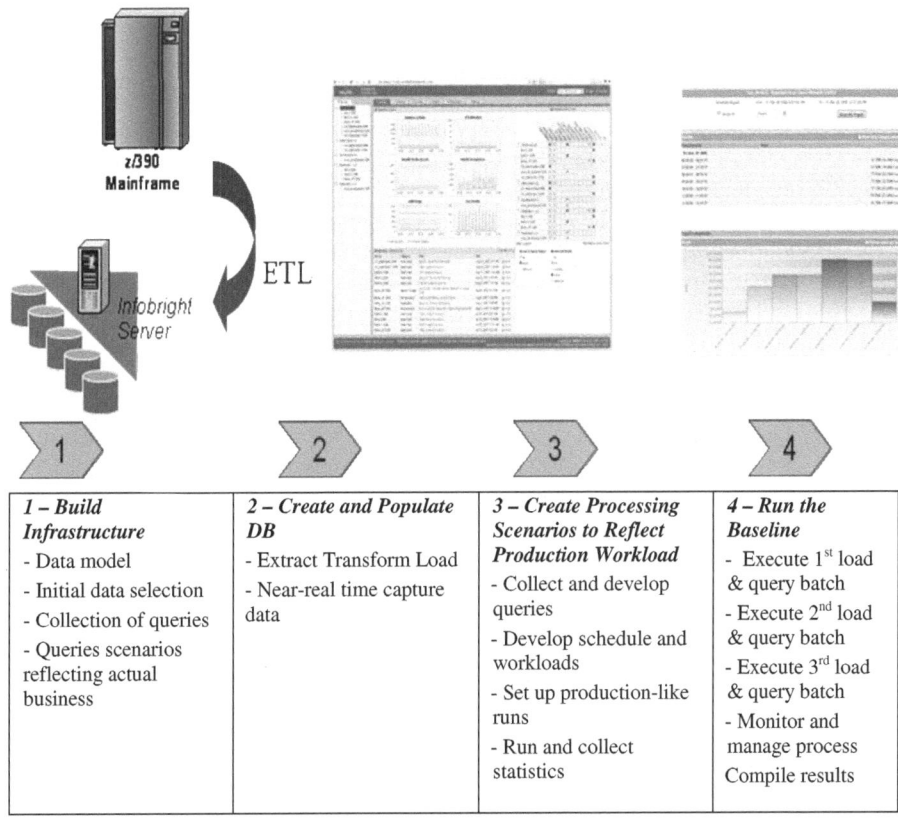

1 – Build Infrastructure	2 – Create and Populate DB	3 – Create Processing Scenarios to Reflect Production Workload	4 – Run the Baseline
- Data model - Initial data selection - Collection of queries - Queries scenarios reflecting actual business	- Extract Transform Load - Near-real time capture data	- Collect and develop queries - Develop schedule and workloads - Set up production-like runs - Run and collect statistics	- Execute 1st load & query batch - Execute 2nd load & query batch - Execute 3rd load & query batch - Monitor and manage process Compile results

Fig. 1. Customizable and Portable Database Model (MODEL)

4.1 Cost Effectiveness Criteria

Due to drastic differences in the cost of platforms, the cost-effectiveness should be considered a separate – equal to performance – category of selection criteria. The weight of cost-effectiveness, however, should be considered only if user requirements are completely satisfied by performance criteria. The rule of selection may be set as follows: If the expected Total Cost of Ownership (TCO) of two platforms is not significantly[2] different and both meet the performance requirements, the better performance score should decide selection. If, however, the cost difference is much

---

[2] At this point, Benchmarking team is considering significant difference at 10% or more. It may be further revised based on the subsequent discussions with identified stakeholders.

more substantial and both platforms pass the minimum performance criteria, then the selection should be decided by the cost factor. The appropriate threshold of cost-effectiveness versus platform performance should be decided during benchmarking as a collective effort. The benchmarking process will produce the TCO calculator that will put together comprehensive calculation of most prominent platforms.

5   Conclusions

Any DW/BI efforts are a very costly and time consuming process. Any design and architectural inaccuracy have usually very profound and long lasting adverse effects. This is the reason that platform selection for the right DW/BI is extremely crucial and should be conducted at the very early stage of such project. This calls for more formal and universal methodology to scrutinize any candidate. We have made an attempt to develop such methodology. The new formal approach was presented: firs we created Preference Matrix that allows company to express specific environmental needs of the organization by assigning more weights to the group of performance factors that is more important to the organization. Secondly, we created the matrix of specific attributes that any candidate platform will be measured against. And lastly a portable and customizable data model and query/load scenario was developed to run a Baseline against. The two matrices, as well as portable model and query/load scenario, are novel concepts formalized based on years of author's experience and best industry practices in the area of DW/BI.

References

1. Infobright IEE: Undertaking a Proof of Concept. Infobright Inc. (2008)
2. Imhoff, C.: There Must Be A Better Way: Building Quick, Inexpensive, and Flexible BI Environments (2008)
3. Darmont, J., Bentayeb, F., Boussaid, O.: DWEB: A Data Warehouse Engineering Benchmark (2007)
4. Aberdeen Group: Is Your Infrastructure Ready to Win RFID?(2008)
5. Aberdeen Group: Aberdeen-IT-architecture-optimization (2008)
6. Government of Alberta: Benchmarking Best Practices (2008)
7. Farrell & Associates Pty.: Benchmarking of IT Services (2002)
8. Harvard Business School: Best Practices for Benchmarking. HBS Working Knowledge (2003)
9. Oracle: Conducting a Data Warehousing Benchmark (2006)
10. Gartner: Server Evaluation Model Summary: Data Warehouse DBMS (2007)
11. Winter Corporation: Advance in Data Warehouse Performance – I/O Elimination in DB2 (2007)
12. A Guide to Lower Database TCO – How the Open Source Database MySQL Reduces Costs by as Much as 90% (2007)
13. Niva, M., Tuominen, K.: Benchmarking in Practice Self-assessment Work Book – Good practices and benchmarking studies. Benchmarking Ltd
14. Apanowicz, C.: Data Warehouse Methodology and Practice. Collection of proprietary materials

Automatic Extraction of Decision Rules from Non-deterministic Data Systems: Theoretical Foundations and SQL-Based Implementation

Dominik Ślęzak[1,2] and Hiroshi Sakai[3]

[1] Institute of Mathematics, University of Warsaw
Banacha 2, 02-097 Warsaw, Poland
[2] Infobright Inc., Poland
Krzywickiego 34 pok. 219, 02-078 Warsaw, Poland
slezak@infobright.com
[3] Department of Basic Sciences,
Faculty of Engineering, Kyushu Institute of Technology
Tobata, Kitakyushu 804, Japan
sakai@mns.kyutech.ac.jp

Abstract. We present a framework for SQL-based extraction of decision rules from data, with no need of retrieving massive amounts of rows from a database. We also explain how to design efficient methods for mining non-deterministic data, without any intermediate stages related to the analysis of undetermined values.

Keywords: Non-Deterministic Information Systems, Data Mining, If-Then Rules, Apriori Algorithm, Relational Databases, RDF, SQL.

1 Introduction

Non-determinism is an important topic in the areas of data mining and data processing. There is a growing interest in uncertain data in the community of database researchers and practitioners (see e.g. the related sessions at [3] or discussions at [7]). There is also a lot of research on theoretical foundations of information systems with uncertain values (see e.g. [5,10]).

The most common case of data uncertainty relates to NULLs. In relational databases, when executing SQL statements like, e.g., *select x,y,z,avg(a) from t group by x,y,z;*, there is a standard way of handling NULLs. Not so many people, however, pay attention to whether it meets real-life expectations. For every group determined by columns x,y,z, there may be rows partially filled in with NULLs that might or might not be regarded as its members, depending on interpretation. Consequently, the user may wish to analyze approximations of $avg(a)$ for particular x,y,z-groups, depending on interpretation of NULLs.

Incompleteness is also a challenge in data mining. Even relatively simple, well-known data mining tasks become complicated and sometimes not so obvious to re-define when the values are not crisply identified. Let us consider an example of

D. Ślęzak et al. (Eds.): DTA 2009, CCIS 64, pp. 151–162, 2009.

extracting association rules [1,2] or, more generally, meaningful IF-THEN rules from data. Although it seems so easy to formulate, once there are missing values involved, we have to decide how to understand the rules' accuracy, support and coverage. In other words, we need to decide when a given data row should be regarded as matching a rule's left/right part [6,9].

As written above, NULLs are just one of examples of uncertainty. Depending on applications, we may need to deal with, e.g., probabilistic, fuzzy or interval values. Yet another example is the notion of a *Non-deterministic Information System (NIS)* [5,10], where, unlike in a *Deterministic Information System (DIS)*, objects (rows, records) can be assigned with subsets of possible values of attributes (columns, features). Comparing to databases with NULLs, NISs may be regarded as a source of incomplete information, rather than incomplete data. In NISs, the specific case of a NULL can be expressed as the full set of values possible for a given attribute. The framework of NISs is, however, far richer.

In this paper, we continue our research on defining and generating IF-THEN rules from NISs [14,15]. Unlike in other approaches, wherein the phase of rule generation is preceded by various statistical, heuristic procedures of completing data/information, we derive the rules directly from non-deterministic data. Furthermore, instead of estimating the rules' support, accuracy, etc., we provide their approximations that hold when reconsidering those rules for arbitrary DISs that might result from the given NIS by completing information it represents. Fortunately, we do not need to explicitly analyze all such hypothetically possible DISs when computing approximations. In our previous research, we show how to derive the rules and their parameters' approximations directly from NISs, without any additional preprocessing stages that would make the whole process too complicated and time-consuming.

The main contribution of this paper lays in redefining the analysis of NISs using SQL, in a form analogous to RDF data stored in a relational database structure [4,17]. We show that SQL-based version of Apriori algorithm [1,2] for decision rules (association rules with the fixed right-hand-side) can be successfully executed in both DISs and NISs. In particular, we show that the above-mentioned NIS-related approximations of rules' parameters are fully expressible in SQL. It proves that the methodology established in [14,15] can be re-implemented within a database framework, which is important especially for the analysis of NISs with large amounts of objects and attributes, using such technologies as, e.g., Infobright Community Edition (ICE) [7,18]. Indeed, it is an ongoing research direction how to redesign the data mining tasks to work basing on intermediate results of dynamically generated SQL aggregations instead of extraction of large amounts of raw data to be processed outside a database.

The paper is organized as follows: Section 2 recalls basic concepts of automatic extraction of IF-THEN rules from data. Section 3 sketches SQL-based methodology for computing decision rules. Section 4 outlines fundamental challenges related to non-deterministic data and gives some solutions with respect to decision rules. Section 5 extends our SQL-based procedures to let it work with non-deterministic data. Section 6 concludes the paper.

Table 1. DIS with $OB = \{1, .., 8\}$ and $AT = \{Temperature, Headache, Nausea, Flu\}$

OB	$Temperature$	$Headache$	$Nausea$	Flu
1	$high$	yes	no	yes
2	$high$	yes	yes	yes
3	$normal$	no	no	yes
4	$high$	yes	yes	yes
5	$high$	yes	yes	no
6	$normal$	yes	yes	yes
7	$normal$	no	yes	no
8	$normal$	yes	yes	yes

2   Decision Rules in Deterministic Information Systems

A *Deterministic Information System (DIS)* is a quadruplet (OB, AT, VAL, f), where OB is the set of *objects*, AT is the set of *attributes*, VAL equals to $\bigcup_{A \in AT} VAL_A$ wherein VAL_A is the set of possible values of $A \in AT$, and $f : OB \times AT \rightarrow VAL$ is the assignment of attributes' values to the objects [13]. We may identify DIS with a relational data table, as illustrated by Table 1.

IF-THEN rules are often applied to express regularities in data. A rule τ is an implication $\tau_{con} \Rightarrow \tau_{dec}$, where τ_{con} and τ_{dec} are conjunctions of descriptors $(A, val) \in AT \times VAL$. An object $x \in OB$ supports (A, val), if $f(x, A) = val$. It supports τ_{con} (or τ_{dec}), if it supports all descriptors in τ_{con} (or τ_{dec}). We denote by $\|(A, val)\|$, $\|\tau_{con}\|$, $\|\tau_{dec}\|$ the sets of objects supporting (A, val), τ_{con}, τ_{dec}. Usually, τ is assigned with its support $sup(\tau) = Card(\|\tau_{con}\| \cap \|\tau_{dec}\|)/card(OB)$ and accuracy $acc(\tau) = Card(\|\tau_{con}\| \cap \|\tau_{dec}\|)/Card(\|\tau_{con}\|)$.

Association rules were originally considered for binary data [1]. However, they can be easily extended towards arbitrary IF-THEN rules in arbitrary DISs. There are plenty of approaches to search for significant association rules [2]. The original Apriori algorithm introduced in [1] aimed at generating all rules satisfying predefined constraints for support and accuracy. The corresponding search problem, parameterized by $\alpha, \beta \in [0, 1]$, is recalled below.

Definition 1. (α, β)-**Association Rule Problem:** *For the given DIS, find all rules τ such that $sup(\tau) \geq \alpha$, $acc(\tau) \geq \beta$, and removing any descriptor from τ_{con} or adding any descriptor to τ_{dec} violates at least one of these inequalities.*

In this paper, we are interested in rules τ, where τ_{dec} is always based on the same attribute, called decision. Such implications, often called decision rules (cf. [6,9]), are especially useful in classification and prediction [8,11]. In the next section, we outline how the following problem, analogous to Definition 1, can be solved using SQL procedures executed against DISs stored in a database.

Definition 2. (α, β)-**Decision Rule Problem:** *For DIS with fixed decision attribute d, find all rules τ with τ_{dec} based only on d and τ_{con} not involving d, such that $sup(\tau) \geq \alpha$, $acc(\tau) \geq \beta$, and removing any descriptor from τ_{con} violates at least one of these inequalities.*

Table 2. DIS_{RDF} for DIS in Table 1. rid stands for an object's identifier.

rid	A	val		rid	A	val		rid	A	val		rid	A	val
1	Temp.	high		3	Temp.	norm.		5	Temp.	high		7	Temp.	norm.
1	Head.	yes		3	Head.	no		5	Head.	yes		7	Head.	no
1	Naus.	no		3	Naus.	no		5	Naus.	yes		7	Naus.	yes
1	Flu	yes		3	Flu	yes		5	Flu	no		7	Flu	no
2	Temp.	high		4	Temp.	high		6	Temp.	norm.		8	Temp.	norm.
2	Head.	yes		4	Head.	yes		6	Head.	yes		8	Head.	yes
2	Naus.	yes		4	Naus.	yes		6	Naus.	yes		8	Naus.	yes
2	Flu	yes		4	Flu	yes		6	Flu	yes		8	Flu	yes

3   Searching for Decision Rules Using SQL

For large data sets stored in data warehouses, there is certainly a question whether it is possible to solve the problems such as those in Definitions 1 and 2 without a need of getting the raw data out. In [17], as an example, SQL-based algorithms computing association rules were discussed. There are also other examples in the literature, e.g., rewriting classical algorithms constructing decision trees [11] to work with dynamically generated SQL aggregate statements [12].

Let us follow the above idea for the case of Definition 2. The proposed procedure is analogous to that in [17]. First of all, we need an alternative approach to storing data. It will turn out helpful also in the non-deterministic case later. It refers to so called RDF data format [4]. In our case, as illustrated by Table 2, it means rewriting a DIS in the form of a triple-store, where each row represents a triplet $(object, attribute, value) \in OB \times AT \times VAL$, consistent with the original assignment f. Let us refer to such a rewritten DIS as DIS_{RDF}.

The reminder of this section outlines how to use SQL to solve (α, β)-Decision Rule Problem for DIS_{RDF}. We describe only the initial steps of the iterative process. Generally, it computes data tables C_i, D_i, F_i, wherein:

- The rows in C_i encode the conjunctions of i descriptors that are satisfied by not less than α times $card(OB)$ objects;
- The rows in D_i encode decision rules τ, where τ_{con} consists of i descriptors, such that $sup(\tau) \geq \alpha$;
- The rows in F_i encode decision rules τ taken from D_i, but such that additionally $acc(\tau) < \beta$;
- The remaining rules encoded by D_i are moved to the set OUTPUT, which will eventually contain all rules holding constraints of Definition 2.

C_{i+1} and D_{i+1} are created based on C_i and F_i. The rows in D_{i+1} that encode rules τ such that $sup(\tau) \geq \alpha$ and $acc(\tau) \geq \beta$ are appended to OUTPUT, while the rest of rows is moved to F_{i+1}. It is easy to see that the rules in OUTPUT are irreducible by means of Definition 2. We stop when F_{i+1} is empty.

Proposition 1. *For the given DIS rewritten as DIS_{RDF}, for the given $\alpha, \beta \in [0, 1]$ and decision attribute d, the above-outlined procedure solves the (α, β)-Decision Rule Problem. The set of all rules that satisfy constraints of Definition 2 corresponds to the set OUTPUT, after we reach the stop criterion.*

Instead of a formal proof, let us sketch a few first steps of the procedure. We start with table D_0 encoding decision values, and table C_1 encoding single-descriptor conditions. We encode descriptors by the pairs of columns. The decision parts are encoded by a single column because $d \in AT$ is fixed anyway.

```
insert into D_0
 select    val, count(*) as sup
 from      DIS_RDF
 where     att = d
 group by val;
```

```
insert into C_1
 select    att, val, count(*) as sup
 from      DIS_RDF
 where     att != d
 group by att, val
 having    sup >= alpha * card(OB);
```

Let us note that the rows in D_0 such that $sup \geq \beta * card(OB)$ are moved directly to OUTPUT as the empty rules. The remaining rows are moved into table F_0. The next step is to create D_1 that stores rules with single conditions:[1]

```
insert into D_1
 select    C_1.att, C_1.val, F_0.val as dec,
           count(*) as sup, count(*)/min(C_1.sup) as acc
 from      C_1, F_0, DIS_RDF t_1, DIS_RDF t_d
 where     t_1.att = C_1.att and t_1.val = C_1.val and
           t_d.att = d and t_d.val = F_0.val and t_1.rid = t_d.rid
 group by C_1.att, C_1.val, F_0.val
 having    sup >= alpha * card(OB);
```

The rules encoded in D_1 are split onto those appended to OUTPUT $(acc \geq \beta)$ and those moved to F_1 $(acc < \beta)$. If F_1 is not empty, the procedure continues with building C_2 and D_2. We finish this section by presenting appropriate SQL statements, without continuing with F_2 and further iterations. It is worth noting that the usage of condition $att_1 < att_2$ when building C_2 (and analogous conditions for C_i, $i > 2$) prevents us from encoding the same conjunctions of descriptors by multiple rows in the same table. Obviously, the whole procedure can be implemented as fully automatic, launched for the given $\alpha, \beta \in [0, 1]$.

---

[1] Everything seems to be intuitive except the usage $min(C_1.sup)$. However, the value of $C_1.sup$ is uniquely defined by $C_1.att$ and $C_1.val$ in the group by clause. We use $min(C_1.sup)$ instead of $C_1.sup$ just to make SQL syntax correct. We might write $C_1.sup$ instead of $min(C_1.sup)$ and it would still work for some database platforms.

Table 3. A Non-deterministic Information System (NIS)

OB	Temperature	Headache	Nausea	Flu
1	{high}	{yes, no}	{no}	{yes}
2	{high, very_high}	{yes}	{yes}	{yes}
3	{normal, high, very_high}	{no}	{no}	{yes, no}
4	{high}	{yes}	{yes, no}	{yes, no}
5	{high}	{yes, no}	{yes}	{no}
6	{normal}	{yes}	{yes, no}	{yes, no}
7	{normal}	{no}	{yes}	{no}
8	{normal, high, very_high}	{yes}	{yes, no}	{yes}

```
insert into C_2
  select   c_1.att as att_1, c_1.val as val_1, c_2.att
           as att_2, c_2.val as val_2, count(*) as sup
  from     C_1 c_1, C_1 c_2, DIS_RDF t_1, DIS_RDF t_2
  where    t_1.att = c_1.att and t_1.val = c_1.val and
           t_2.att = c_2.att and t_2.val = c_2.val and
           t_1.rid = t_2.rid and att_1 < att_2
  group by c_1.att, c_1.val, c_2.att, c_2.val
  having   sup >= alpha * card(OB);

insert into D_2
  select   C_2.att_1, C_2.val_1, C_2.att_2, C_2.val_2, f_1.val
           as dec, count(*) as sup, count(*)/min(C_2.sup) as acc
  from     C_2, F_1 f_1, F_1 f_2,
           DIS_RDF t_1, DIS_RDF t_2, DIS_RDF t_d
  where    C_2.att_1 = f_1.att and C_2.val_1 = f_1.val and
           C_2.att_2 = f_2.att and C_2.val_2 = f_2.val and
           t_1.att = C_2.att_1 and t_1.val = C_2.val_1 and
           t_2.att = C_2.att_2 and t_2.val = C_2.val_2 and
           t_d.att = d and t_d.val = f_1.dec and t_d.val =
           f_2.dec and t_1.rid = t_d.rid and t_2.rid = t_d.rid
  group by C_2.att_1, C_2.val_1, C_2.att_2, C_2.val_2, f_1.val
  having   sup >= alpha * card(OB);
```

4   Rules in Non-deterministic Information Systems

A *Non-deterministic Information System (NIS)* [5,10] is a quadruplet (OB, AT, VAL, g). Comparing to DISs, f is replaced by $g : OB \times AT \rightarrow P(VAL)$. Table 3 shows an example of NIS. Every $g(x, A) \subseteq VAL_A$ is interpreted as that there is an actual value of x on A but, for some real-life reason, it is not fully determined. One may imagine multiple hypothetical DISs that might be hidden behind non-deterministic information represented by a given NIS.

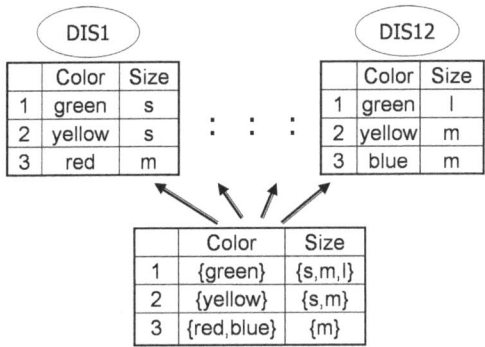

Fig. 1. An example of DISs that can be derived from a given NIS

Definition 3. *Let NIS (OB, AT, VAL, g) and DIS (OB, AT, VAL, f) be given. We say that DIS is derived from NIS, if and only if there is $f(x, A) \in g(x, A)$ for each $x \in OB, A \in AT$.*

Fig. 1 illustrates how many DISs can be derived from a single NIS. In the areas of data mining and machine learning, one has a choice to redesign standard approaches to let them work with NISs, or to develop an additional preprocessing stage of replacing a NIS with one of derived DISs. That latter scenario is quite popular in the analysis of data with missing values, where various techniques are applied to replace them with the most probable elements of *VAL*. One may say that this is the way of constructing the most probable DIS, which can be then an input to a standard algorithm searching for association rules, decision trees, et cetera. However, if the amount of missing values is too large, such preprocessing becomes not reliable enough – application of an inaccurate method of filling in the missing values may result in misleading results of the data mining algorithms at the next stage. In the same way, if the cardinalities of sets $g(x, A) \subseteq VAL_A$ in a NIS are too large on average, then the number of derived DISs grows very fast and the choice of a single *best* derived DIS becomes impossible.

Another strategy would be to work with all (or a sample of) DISs derived from the given NIS. For example, for a given rule τ, one may search for derived DISs, such that $sup(\tau)$ and $acc(\tau)$ are minimal and maximal. Such obtained quantities, denoted by $minsup(\tau)$, $minacc(\tau)$, $maxsup(\tau)$, and $maxacc(\tau)$, respectively, may be then treated as *approximations* of τ's parameters in the given NIS. For example, if someone wants to search for decision rules, such that support and accuracy are good enough no matter what the actual values in the given NIS might turn out to be, Definition 2 should be rewritten as follows:

Definition 4. (α, β)**-Decision Rule Problem (Non-deterministic Data):** *For the given NIS with the fixed decision attribute d, find all implications τ with τ_{dec} based only on d and τ_{con} not involving d, satisfying $minsup(\tau) \geq \alpha$ and $minacc(\tau) \geq \beta$, and such that removing any component from τ_{con} would violate at least one of these inequalities.*

Obviously, it would be extremely difficult to solve such problems, if we had to operate explicitly on multiple DISs derived from the given NIS. In [15], it was shown that the quantities of $minsup(\tau)$ and $minacc(\tau)$ (or $maxsup(\tau)$ and $maxacc(\tau)$)) can be computed over the same derived DIS, denoted as DIS^τ_{worst} (or DIS^τ_{best}). Such results help in understanding the complexity of the problem. However, it is required to design such techniques of computing (or at least estimating) the quantities of $minsup$, $minacc$, $maxsup$, $maxacc$, as well as such algorithms solving the problems similar to the one introduced in Definition 4 that would be able to work *directly* on the data represented by NISs, without a need of generating multiple intermediate DISs.

In [15], several methods for computing the above quantities directly from NISs were reported. Because of a lack of space, let us describe just one of them. We refer to the previous publications for details and formal proofs. Let us start by generalizing the concept of supporting descriptors by objects. For $A \in AT$ and $val \in VAL_A$, consider the following sets:

$$
\begin{aligned}
\|(A, val)\|_{all} &= \{x \in OB | g(x, A) = \{val\}\} \\
\|(A, val)\|_{som} &= \{x \in OB | val \in g(x, A)\}
\end{aligned}
\tag{1}
$$

The set $\|(A, val)\|_{all}$ (or $\|(A, val)\|_{som}$) contains all objects that would support (A, val) in all (or at least one of) DISs derived from a given NIS. Analogously, we can introduce the quantities of $\|\tau_{con}\|_{all}$, $\|\tau_{dec}\|_{all}$, $\|\tau_{con}\|_{som}$ and $\|\tau_{dec}\|_{som}$. The following equations, proved as valid in [15], enable us to compute the quantities of $minsup(\tau)$ and $minsup(\tau)$ that are considered in Definition 4:

$$
\begin{aligned}
minsup(\tau) &= Card(\|\tau_{con}\|_{all} \cap \|\tau_{dec}\|_{all}) / Card(OB) \\
minacc(\tau) &= Card(\|\tau_{con}\|_{all} \cap \|\tau_{dec}\|_{all}) / [Card(\|\tau_{con}\|_{all}) \\
&\quad + Card((\|\tau_{con}\|_{som} \setminus \|\tau_{con}\|_{all}) \setminus \|\tau_{dec}\|_{all})]
\end{aligned}
\tag{2}
$$

Analogous equations can be formulated for $maxsup(\tau)$ and $maxacc(\tau)$ as well, in case someone wants to reconsider the (α, β)-Decision Rule Problem for the rules that might satisfy the constraints for some (but not necessarily all) of possible DISs derived from the given NIS:

$$
\begin{aligned}
maxsup(\tau) &= Card(\|\tau_{con}\|_{som} \cap \|\tau_{dec}\|_{som}) / Card(OB) \\
maxacc(\tau) &= [Card(\|\tau_{con}\|_{all} \cap \|\tau_{dec}\|_{sup}) \\
&\quad + Card((\|\tau_{con}\|_{som} \setminus \|\tau_{con}\|_{all}) \cap \|\tau_{dec}\|_{all})] \\
&\quad / [Card(\|\tau_{con}\|_{all}) \\
&\quad + Card((\|\tau_{con}\|_{som} \setminus \|\tau_{con}\|_{all}) \cap \|\tau_{dec}\|_{all})]
\end{aligned}
\tag{3}
$$

The authors of [15] implemented an extended version of Apriori algorithm, called *NIS-Apriori*, that iteratively scans through non-deterministic data in order to search for rules τ satisfying various constraints defined by means of $minsup(\tau)$, $minacc(\tau)$, $maxsup(\tau)$ and $maxacc(\tau)$. The experiments show that, thanks to the usage of equations such as (2,3), NIS-Apriori is far more efficient than any techniques attempting to construct intermediate DISs from NISs. In this paper, we go a step further and, given motivation introduced in Sections 1 and 3, we focus on rewriting NIS-Apriori in terms of SQL.

Table 4. A NIS in RDF format. Column *det* indicates whether values are determined.

rid	A	val	det
1	Temp.	high	1
1	Head.	yes	0
1	Head.	no	0
1	Naus.	no	1
1	Flu	yes	1
2	Temp.	high	0
2	Temp.	v._h.	0
2	Head.	yes	1
2	Naus.	yes	1
2	Flu	yes	1
3	Temp.	norm.	0
3	Temp.	high	0
3	Temp.	v._h.	0
3	Head.	no	1
3	Naus.	no	1

rid	A	val	det
3	Flu	yes	0
3	Flu	no	0
4	Temp.	high	1
4	Head.	yes	1
4	Naus.	yes	0
4	Naus.	no	0
4	Flu	yes	0
4	Flu	no	0
5	Temp.	high	1
5	Head.	yes	0
5	Head.	no	0
5	Naus.	yes	1
5	Flu	no	1
6	Temp.	norm.	1
6	Head.	yes	1

rid	A	val	det
6	Naus.	yes	0
6	Naus.	no	0
6	Flu	yes	0
6	Flu	no	0
7	Temp.	norm.	1
7	Head.	no	1
7	Naus.	yes	1
7	Flu	no	1
8	Temp.	norm.	0
8	Temp.	high	0
8	Temp.	v._h.	0
8	Head.	yes	1
8	Naus.	yes	0
8	Naus.	no	0
8	Flu	yes	1

5 Searching for Decision Rules in NISs – SQL Perspective

The main purpose of this part of the paper is to show that the procedure described in Section 3 can be fully adapted to searching for decision rules in NISs. As a case study, we focus on solving the problem stated in Definition 4 and we work with equations (2). However, the analogous procedures can be introduced also for other versions of the (α, β)-Decision Rule Problem in NISs.

Just like before, we use the data format analogous to RDF. Its additional advantage with respect to NISs (OB, AT, VAL, g) is that it can easily represent multiple values stored in sets $g(x, A)$ as separate rows. Table 4 illustrates how it is done for a NIS from Table 3. We denote such a data format as NIS_{RDF}.

An important addition to NIS_{RDF} is the column *det* that indicates whether particular sets $g(x, A)$ are singletons. It helps in more convenient SQL-based computations of equations (1). Certainly, one might derive the values of *det* dynamically, without a need of extending the data model. We decided to physically add *det* to NIS_{RDF} in order to simplify SQL. The values of *det* can be computed once, when transforming original NISs to the NIS_{RDF} format. Such operation is independent from a choice of parameters $\alpha, \beta \in [0, 1]$ or decision d in the specification of particular instance of the search problem.

The implemented procedure is indeed fully analogous to the one outlined for DISs in Section 3. Just like before, it dynamically creates data tables C_i, D_i, F_i and stores rules that are components of the final result in the set OUTPUT. Just like before, it can be proved that OUTPUT represents the solution of the problem formulated in Definition 4, when the algorithm reaches the stop criterion. The only difference is that now we need to use SQL to compute the quantities of $minsup(\tau)$ and $minacc(\tau)$, instead of $sup(\tau)$ and $acc(\tau)$.

Again, because of a lack of space, we provide a few examples of SQL statements that are automatically generated during the considered procedure. Assume that tables C_1 and F_1 are already in place and that F_1 is not empty. Below we present SQL for C_2 and D_2. The case of C_2 is almost identical to the analogous statement in Section 3. The only difference is that now we need to count the elements of $\|\tau_{con}\|_{all}$, so we need additional filters $t_1.det = 1$ and $t_2.det = 1$.

```
insert into C_2
  select    c_1.att as att_1, c_1.val as val_1, c_2.att
            as att_2, c_2.val as val_2, count(*) as minsup
  from      C_1 c_1, C_1 c_2, DIS_RDF t_1, DIS_RDF t_2
  where     t_1.att = c_1.att and t_1.val = c_1.val and
            t_2.att = c_2.att and t_2.val = c_2.val and
            t_1.rid = t_2.rid and att_1 < att_2 and
            t_1.det = 1 and t_2.det = 1
  group by  c_1.att, c_1.val, c_2.att, c_2.val
  having    minsup >= alpha * card(OB);
```

The case of D_2 is slightly more complicated because it requires operating with both $\|\cdot\|_{all}$ and $\|\cdot\|_{som}$ versions of the support defined by equations (1). In the following SQL statement, the aggregate function $sum(min(t_1.det, t_2.det, t_d.det))$ is responsible for computing cardinalities $Card(\|\tau_{con}\|_{all} \cap \|\tau_{dec}\|_{all})$ for the produced decision rules τ with two descriptors at their left-hand-sides. The component $min(C_2.minsup)$ (or $C_2.minsup$ – see the footnote in Section 3) corresponds to $Card(\|\tau_{con}\|_{all})$. Finally, the component $sum(max(t_1.det, t_2.det) - max(min(t_1.det, t_2.det), t_d.det))$ completes the way of expressing $minacc(\tau)$ by computing $Card((\|\tau_{con}\|_{som} \setminus \|\tau_{con}\|_{all}) \setminus \|\tau_{dec}\|_{all})$. Just like in case of DISs in Section 3, the rules τ corresponding to rows in D_2 are then split with respect to satisfaction of criterion $minacc(\tau) \geq \beta$. If the resulting F_2 is not empty, the algorithm continues with appropriate SQL statements for $C_i, D_i, i > 2$.

```
insert into D_2
  select    C_2.att_1, C_2.val_1, C_2.att_2, C_2.val_2, f_1.val
            as dec, sum(min(t_1.det,t_2.det,t_d.det)) as minsup,
            sum(min(t_1.det,t_2.det,t_d.det))/(min(C_2.minsup) +
            sum(max(t_1.det,t_2.det) - max(min(t_1.det,t_2.det),
            t_d.det))) as minacc
  from      C_2, F_1 f_1, F_1 f_2,
            DIS_RDF t_1, DIS_RDF t_2, DIS_RDF t_d
  where     C_2.att_1 = f_1.att and C_2.val_1 = f_1.val and
            C_2.att_2 = f_2.att and C_2.val_2 = f_2.val and
            t_1.att = C_2.att_1 and t_1.val = C_2.val_1 and
            t_2.att = C_2.att_2 and t_2.val = C_2.val_2 and
            t_d.att = d and t_d.val = f_1.dec and t_d.val =
            f_2.dec and t_1.rid = t_d.rid and t_2.rid = t_d.rid
  group by  C_2.att_1, C_2.val_1, C_2.att_2, C_2.val_2, f_1.val
  having    minsup >= alpha * card(OB);
```

6   Concluding Remarks

We introduced a framework for SQL-based extraction of decision rules from DISs and NISs. On the one hand, we referred to the previous research related to a usage of SQL and RDF format in data mining [4,17]. On the other hand, we followed our own research on representing and finding optimal IF-THEN rules in non-deterministic data [14,15]. The obtained results illustrate that standard SQL can be applied in complex analytic tasks. As pointed out in Section 1, redesigning such tasks to work to a larger extent with SQL aggregations is especially important in case of massive data. It should be also emphasized one more time that analyzing data sets with not fully determined values requires a thorough study, far beyond the current standards of interpreting NULLs in SQL or estimating missing values in machine learning.

The future research directions are threefold. The first one is performance. We based our initial experiments on Infobright Community Edition (ICE) [7,18]. A helpful feature of ICE is that it does not need classical database indices, which would be quite useless for SQL in Sections 3 and 5. Moreover, it enables to work with terabytes of data on standard equipment. However, the proposed SQL statements may still need improvements. The speed may also relate to specific types of algorithms' outcomes. For example, one may construct classification/prediction models using only a subset of most representative rules [8]. On the other hand, one may obtain even larger amounts of rules, if the constraint for a rule's left-hand-side irreducibility is not in place. All such changes in requirements may certainly influence performance.

The second direction is interpretation of rules' parameters. We showed that SQL can be efficiently applied to compute approximations of the rules' support and accuracy (the same could be done for the rules' coverage; see [7,15]) over hypothetical DISs that might be derived from the given NIS. However, one may be interested in searching for decision rules with less strict constraints for support and accuracy than those in Definition 4. For instance, one may wish to work with decision rules τ such that inequalities $sup(\tau) \geq \alpha$ and $acc(\tau) \geq \beta$ are satisfied for *almost* all derived DISs. Mathematical foundations for the corresponding search problems are provided in [14]. It is an open question whether replacing binary det by a column with the values of the form $1/Card(g(x, A))$ in NIS_{RDF} may help to extend the SQL framework accordingly.

The third direction is to deal with more sophisticated types of data uncertainty (cf. [3,7]). As an example, we are now investigating whether the framework described in this paper can be extended towards values in form of numeric intervals. On the one hand, it needs an interval-based analogy of Definition 3. On the other hand, it requires changes in SQL and in NIS_{RDF} format. A separate issue is how to generalize descriptors that are atomic components of constructed decision rules. In case of numeric data, descriptors can be based on inequalities or intervals. Furthermore, even in case of symbolic data, descriptors can be specified as subsets, instead of single elements of VAL_A (cf. [10,16]). It shows that extensions of the expressive power of our approach should correspond to both new types of data and new types of decision rules.

Acknowledgements. The first author was partially supported by the grants N N516 368334 and N N516 077837 from the Ministry of Science and Higher Education of the Republic of Poland. The second author was partially supported by the Grant-in-Aid for Scientific Research (C) (No. 16500176, No. 18500214) from the Japan Society for the Promotion of Science.

References

1. Agrawal, R., Srikant, R.: Fast Algorithms for Mining Association Rules. In: Proc. of VLDB, pp. 487–499 (1994)
2. Ceglar, A., Roddick, J.F.: Association mining. ACM Comput. Surv. 38(2) (2006)
3. Çetintemel, U., Zdonik, S.B., Kossmann, D., Tatbul, N. (eds.) Proc. of SIGMOD (2009)
4. Chong, E.I., Das, S., Eadon, G., Srinivasan, J.: An efficient SQL-based RDF querying scheme. In: Proc. of VLDB, pp. 1216–1227 (2005)
5. Demri, S., Orłowska, E.: Incomplete Information: Structure, Inference, Complexity. Springer, Heidelberg (2002)
6. Grzymała-Busse, J.: Data with Missing Attribute Values: Generalization of Indiscernibility Relation and Rule Induction. Trans. Rough Sets 1, 78–95 (2004)
7. Infobright.org Forums, http://www.infobright.org/Forums/viewthread/288/, http://www.infobright.org/Forums/viewthread/621/
8. Kloesgen, W., Żytkow, J.M. (eds.): Handbook of Data Mining and Knowledge Discovery. Oxford University Press, Oxford (2002)
9. Kryszkiewicz, M.: Rules in Incomplete Information Systems. Information Sciences 113, 271–292 (1999)
10. Lipski, W.: On Semantic Issues Connected with Incomplete Information Data Base. ACM Trans. DBS 4, 269–296 (1979)
11. Mitchell, T.: Machine Learning. Mc Graw Hill, Newyork (1998)
12. Nguyen, H.S., Nguyen, S.H.: Fast split selection method and its application in decision tree construction from large databases. Int. J. Hybrid Intell. Syst. 2(2), 149–160 (2005)
13. Pawlak, Z.: Information systems theoretical foundations. Inf. Syst. 6(3), 205–218 (1981)
14. Sakai, H., Hayashi, K., Nakata, M., Ślęzak, D.: The Lower System, the Upper System and Rules with Stability Factor in Non-deterministic Information Systems. In: Proc. of RSFDGrC, LNCS, vol. 5908, pp. 313–320, Springer, Heidelberg (2009) (in print)
15. Sakai, H., Ishibashi, R., Nakata, M.: On Rules and Apriori Algorithm in Non-deterministic Information Systems. Trans. Rough Sets 9, 328–350 (2008)
16. Shafer, G.: A Mathematical Theory of Evidence. Princeton University Press, Princeton (1976)
17. Sarawagi, S., Thomas, S., Agrawal, R.: Integrating Association Rule Mining with Relational Database Systems: Alternatives and Implications. Research Report, http://www.almaden.ibm.com/cs/projects/iis/hdb/Publications/papers/sigmod98_dbi_rj.pdf
18. Ślęzak, D., Eastwood, V.: Data warehouse technology by Infobright. In: Proc. of SIGMOD 2009, pp. 841–845 (2009)

Soft Set Approach for Maximal Association Rules Mining

Tutut Herawan[1,2], Iwan Tri Riyadi Yanto[1], and Mustafa Mat Deris[1]

[1] FTMM, Universiti Tun Hussein Onn Malaysia, Johor, Malaysia
[2] CIRNOV, Universitas Ahmad Dahlan, Yogyakarta, Indonesia
tutut81@uad.ac.id, iwan015@gmail.com, mmustafa@uthm.edu.my

Abstract. In this paper, an alternative approach for maximal association rules mining from a transactional database using soft set theory is proposed. The first step of the proposed approach is based on representing a transactional database as a soft set. Based on the soft set, the notion of items co-occurrence in a transaction can be defined. The definitions of soft maximal association rules, maximal support and maximal confidence are presented using the concept of items co-occurrence. It is shown that by using soft set theory, maximal rules discovered are identical and faster as compared to traditional maximal and rough maximal association rules approaches.

Keywords: Data mining; Maximal association rules; Soft set theory.

1 Introduction

Association rule is one of the most popular data mining techniques and has received considerable attention, particularly since the publication of the AIS and Apriori algorithms [1,2]. Let $I = \{i_1, i_2, \cdots, i_{|A|}\}$, for $|A| > 0$, where $|X|$ is the cardinality of X. refers to the set of literals called *set of items* and the set $D = \{t_1, t_2, \cdots, t_{|U|}\}$, for $|U| > 0$ refers to the database of transactions, where each transaction $t \in D$ is a list of distinct items $t = \{i_1, i_2, \cdots, i_{|M|}\}$, $1 \le |M| \le |A|$ and each transaction can be identified by a distinct identifier *TID*. Let, a set $X \subseteq t \subseteq I$ called an *itemset*. Formally, an *association rule* between sets X and Y is an implication of the form $X \Rightarrow Y$, where $X \cap Y = \phi$. The itemsets X and Y are called *antecedent* and *consequent*, respectively. The *support* of an association rule is defined as a number of transactions in D contain $X \cup Y$ and the *confidence* of an association rule is defined as a ratio of the numbers of transactions in D contain $X \cup Y$ to the number of transactions in D contain X. One of the association rules mining methods is the maximal association rule which is introduced by Feldman *et al.* [3,4,5]. While regular association rules [1] are based on the notion of frequent itemsets which appears in many records, maximal association rules are based on frequent maximal itemsets which appears maximally in many records [6]. Bi *et al.* [7] and Guan *et al.* [6,8] proposed the same approach for discovering maximal association rules using rough set theory [9,10,11]. Their proposed approach is based on a partition on the set of all attributes in a transactional

D. Ślęzak et al. (Eds.): DTA 2009, CCIS 64, pp. 163–170, 2009.
© Springer-Verlag Berlin Heidelberg 2009

database so-called a taxonomy and category of items. Soft set theory [12], proposed by Molodtsov in 1999, is a new general method for dealing with uncertain data. Since the "standard" soft set deals with a Boolean-valued information system, then a transactional database can be represented as a soft set. Inspired from the fact that every rough set can be considered as a soft set [13], in this paper, we present the notion of soft maximal association rules. We define the maximal support and maximal confidence of the maximal association rule based on the maximal co-occurrences of parameters concept in a transactional database under soft set theory. Three main contributions of this work are as follows: Firstly, we present a transformation of a transactional data into a soft set. Secondly, we present the applicability of the soft set theory for maximal association rules mining. Lastly, we show that by using soft set theory, rules discovered are identical and faster to traditional maximal association rules in [3,4,5] and rough maximal association rules in [6,7,8]. The rest of this paper is organized as follows. Section 2 describes fundamental concept of soft set theory. Section 3 describes soft set approach for maximal association rules mining. Finally, the conclusion of this work is described in section 4.

2   Soft Set Theory

Definition 1. (See [13].) *Let E be a set of parameters. A pair* (F, E) *is called a soft set over a universe U, where F is a mapping given by* $F : A \rightarrow P(U)$, *where* $P(U)$ *is the power set of U.*

In other words, a soft set over U is a parameterized family of subsets of the universe U. For $\varepsilon \in A$, $F(\varepsilon)$ may be considered as the set of ε -elements of the soft set (F, A) or as the set of ε -approximate elements of the soft set. Clearly, a soft set is not a (crisp) set.

3   Soft Set Approach for Maximal Association Rules Mining

In this section we present the applicability of soft set theory for maximal association rules mining.

3.1   Transformation of a Transactional Database into a Soft Set

The pre-requisite of the proposed approach is the transactional database need to be transformed into a soft set, where each item is regarded as a parameter. For such transformation, we firstly present the relation between a soft set and a Boolean-valued information system as given in the following proposition.

Proposition 2. *If* (F, E) *is a soft set over the universe U, then* (F, E) *is a binary-valued information system* $S = (U, A, V_{\{0,1\}}, f)$.

Proof. Let (F,E) be a soft set over the universe U, we define a collection of mappings $F = \{f_1, f_2, \cdots, f_n\}$, where

$$f_i : U \to V_1 \text{ and } f_i(x) = \begin{cases} 1, & x \in F(ei_i) \\ 0, & x \notin F(e_i) \end{cases}, \text{ for } i = 1, \cdots, n.$$

Hence, if $A = E$, $V = \bigcup_{e_i \in A} V_{e_i}$, where $V_{e_i} = \{0,1\}$, then a soft set (F,E) can be considered as a binary-valued information system $S = (U, A, V_{\{0,1\}}, f)$.   □

For a transactional database, it is known from that a Boolean-valued information system is an efficient data structure for storing information. This structure is equivalent and efficient, because the column-address of a Boolean-valued attribute doesn't change. Therefore it is possible to check very fast each tuple, by direct access for the appropriate column, whether the searched item is or is not available in the corresponding tuple of the original database. If it appear, then there is a '1', otherwise a '0'. Thus, based on Proposition 2, a transactional data can be represented as a soft set. Throughout this point forward, the pair (F,E) refers to the soft set over the universe U representing a transactional database, $D = \{t_1, t_2, \cdots, t_{|U|}\}$.

3.2 Taxonomy and Categorization Using Soft Set Theory

Based on transformation in the previous sub-section, we present the notions of taxonomy and categorization of items using soft set theory as follows.

Let (F,E) be a soft set over the universe U. A *taxonomy* T of E is a partition of E into disjoint subsets, i.e., $T = \{E_1, E_2, E_3, \cdots, E_n\}$. Each member of T is called a *category*. For an item i, we denote $T(i)$ the category that contain i. Similarly, if X is an itemset all of which are from a single category, then we denote this category by $T(X)$.

Definition 3. *Let (F,E) be a soft set over the universe U and $u \in U$. An items co-occurrence set in a transaction u can be defined as $\text{Coo}(u) = \{e \in E : f(u,e) = 1\}$.*

3.3 Soft Maximal Association Rules

In the proposed approach, we use the notion of co-occurrence of parameters for association rules mining as used in [7].

Definition 4. *Let (F,E) be a soft set over the universe U and $X \subseteq E_i$. A set of attributes X is said to be maximal supported by a transaction u if $X = \text{Coo}(u) \cap E_i$.*

Definition 5. *Let* (F,E) *be a soft set over the universe U and* $X \subseteq E_i$. *The maximal support of a set of parameters X, denoted by* $\sup(X)$ *is defined by the number of transactions U maximal supporting X, i.e.* $M \sup(X) = |\{u : X = \mathrm{Coo}(u) \cap E_i\}|$.

Definition 6. *Let* (F,E) *be a soft set over the universe U and two maximal itemsets* $X,Y \subseteq E_i$, *where* $X \cap Y = \phi$. *A maximal association rule between X and Y is an implication of the form* $X \overset{max}{\Rightarrow} Y$. *The itemsets X and Y are called maximal antecedent and maximal consequent, respectively.*

Definition 7. *Let* (F,E) *be a soft set over the universe U and two maximal itemsets* $X,Y \subseteq E_i$, *where* $X \cap Y = \phi$. *The maximal support of a maximal association rule* $X \Rightarrow Y$, *denoted by* $M \sup\left(X \overset{max}{\Rightarrow} Y \right)$ *is defined by*

$$M \sup\left(X \overset{max}{\Rightarrow} Y \right) = M \sup(X \cup Y) = |\{u : X \cup Y = \mathrm{Coo}(u) \cap E_i\}|$$

Definition 8. *Let* (F,E) *be a soft set over the universe U and two maximal itemsets* $X,Y \subseteq E_i$, *where* $X \cap Y = \phi$. *The confidence of a maximal association rule* $X \overset{max}{\Rightarrow} Y$, *denoted respectively by* $M\mathrm{conf}\left(X \overset{max}{\Rightarrow} Y \right)$ *and* $\mathrm{conf}(X \Rightarrow Y)$ *is defined by*

$$M\mathrm{conf}\left(X \overset{max}{\Rightarrow} Y \right) = \frac{M \sup(X \cup Y)}{M \sup(X)} = \frac{|\{u : X \cup Y = \mathrm{Coo}(u) \cap E_i\}|}{|\{u : X = \mathrm{Coo}(u) \cap E_i\}|}.$$

3.4 Experimental Results

We elaborate the proposed approach through two datasets derived from [14] and [15]. They are implemented in MATLAB version 7.6.0.324 (R2008a). They are executed sequentially on a processor Intel Core 2 Duo CPUs. The total main memory is 1G and the operating system is Windows XP Professional SP3.

a. A dataset derived from the widely used Reuters-21578 [14], a labeled document collection, i.e. a benchmark for text categorization, as follows. Assume that there are 10 articles regarding product corn which relate to the countries USA and Canada and 20 other articles concerning product fish and the countries USA, Canada and France. The soft set representing the dataset is given as $(F,E) = \{\mathrm{USA} = \{1,\cdots,30\}, \mathrm{Canada} = \{1,\cdots,30\}, \mathrm{France} = \{11,\cdots,30\}, \mathrm{corn} = \{1,\cdots,10\}, \mathrm{fish} = \{11,\cdots,30\}\}$. From the soft set, we have the co-occurrence, $\mathrm{Coo}(u_1) = \cdots = \mathrm{Coo}(u_{10}) = \{\mathrm{USA}, \mathrm{Canada}, \mathrm{corn}\}$ and $\mathrm{Coo}(u_{11}) = \cdots = \mathrm{Coo}(u_{30}) = \{\mathrm{USA}, \mathrm{Canada}, \mathrm{France}, \mathrm{fish}\}$. Based on the co-occurrences, we can create a taxonomy, $T = \{\mathrm{countries}, \mathrm{products}\}$, where countries = USA, Canada France} and topics = {corn, fish}. The maximal supported sets are given as follows

USACanada	10
USACanadaFrance	20
Corn	10
Fish	20

Fig. 1. The maximal supported sets

Antecedent	Consequent	Msup	Mconf
USACanada	Corn	10	100%
USACanadaFrance	Fish	20	100%
Corn	USACanada	10	100%
Fish	USACanadaFrance	20	100%

Fig. 2. The maximal association rules

The maximal rules captured as in Figure 2 are equivalent with that in [3,6,7,8]. Since on this dataset is easy to capture the maximal rules, we do no comparison in term of executing time. The executing time of the proposed model through this dataset is 0.014 second.

b. We will further explain an example to capture maximal rules based on the observation of the air pollution data taken in Kuala Lumpur on July 2002 as presented and used in [15]. The soft set representing the transactional dataset is given as follows.

$$(F,E) = \begin{cases} a = \{u_1, u_2, u_3, u_4, u_5, u_6, u_7, u_8, u_9, u_{10}, u_{11}, u_{12}, u_{14}, u_{15}, u_{16}, u_{17}, \\ u_{18}, u_{19}, u_{20}, u_{21}, u_{22}, u_{23}, u_{24}, u_{25}, u_{27}, u_{28}, u_{29}, u_{30}\}, b = \{u_6, u_7, u_{18}\}, \\ c = \{u_5, u_6, u_7, u_8, u_{13}, u_{14}, u_{15}, u_{16}, u_{17}, u_{18}, u_{19}, u_{20}, u_{21}, u_{22}, u_{28}, u_{29}\}, \\ d = \{u_3, u_5, u_6, u_7, u_9, u_{10}, u_{11}, u_{12}, u_{13}, u_{14}, u_{16}, u_{17}, u_{18}, u_{19}, u_{20}, u_{21}, u_{22}\}, \\ e = \{u_1, u_5, u_6, u_7, u_8, u_{10}, u_{11}, u_{14}, u_{15}, u_{16}, u_{17}, u_{18}, u_{19}, u_{20}, u_{21}, u_{22}, u_{23}, u_{25}, u_{27}, u_{28}, u_{29}\} \end{cases}$$

Fig. 3. The soft set representing the transaction dataset

From the soft set in Figure 3, we have the following co-occurrences in each transaction.

$\text{Coo}(u_1) = \{a,e\}, \ \text{Coo}(u_2) = \{a\}, \ \text{Coo}(u_3) = \{a,d\}, \ \text{Coo}(u_4) = \{a\},$
$\text{Coo}(u_5) = \{a,c,d,e\}, \ \text{Coo}(u_6) = \{a,b,c,d,e\}, \ \text{Coo}(u_7) = \{a,b,c,d,e\},$
$\text{Coo}(u_8) = \{a,c,e\}, \ \text{Coo}(u_9) = \{a,d,e\}, \ \text{Coo}(u_{10}) = \{a,d,e\}, \ \text{Coo}(u_{11}) = \{a,d,e\},$
$\text{Coo}(u_{12}) = \{a,d\}, \ \text{Coo}(u_{13}) = \{d\}, \ \text{Coo}(u_{14}) = \{a,c,d,e\}, \ \text{Coo}(u_{15}) = \{a,c,e\},$
$\text{Coo}(u_{16}) = \{a,c,d,e\}, \ \text{Coo}(u_{17}) = \{a,c,d,e\}, \ \text{Coo}(u_{18}) = \{a,b,c,d,e\},$
$\text{Coo}(u_{19}) = \{a,c,d,e\}, \ \text{Coo}(u_{20}) = \{a,c,d,e\}, \ \text{Coo}(u_{21}) = \{a,c,d,e\},$
$\text{Coo}(u_{22}) = \{a,c,d,e\}, \ \text{Coo}(u_{24}) = \{a\}, \ \text{Coo}(u_{25}) = \{a,d,e\}, \ \text{Coo}(u_{26}) = \{d\},$
$\text{Coo}(u_{27}) = \{a,d,e\}, \ \text{Coo}(u_{28}) = \{a,c,d,e\}, \ \text{Coo}(u_{29}) = \{a,c,d,e\}, \ \text{Coo}(u_{30}) = \{a,d\}$

Fig. 4. The co-occurrence of items in each transaction

From the transactional dataset, we can create a taxonomy as follows $T = \{\text{dangerous condition, good condition}\}$, where dangerous condition $= \{a, c, d, e\}$ and good condition $= \{b\}$. The maximal supported itemsets are given as follows.

a	2
b	3
ad	2
ae	2
dc	1
ade	4
acde	13

Fig. 5. The maximal supported sets

For capturing interesting maximal rules in the air pollution database, we set the minimum Msupport and minimum Mconfidence as $\min M \sup = 2$ and $\min M \text{conf} = 50\%$, respectively. And the rule discovered is given in Figure 6.

Antecedent	Consequent	Msup	Mconf
b	acde	3	100%

Fig. 6. Maximal association rules obtained

The comparison of executing time of traditional (A), rough (B) and soft (C) maximal association rules is given in the following Figure.

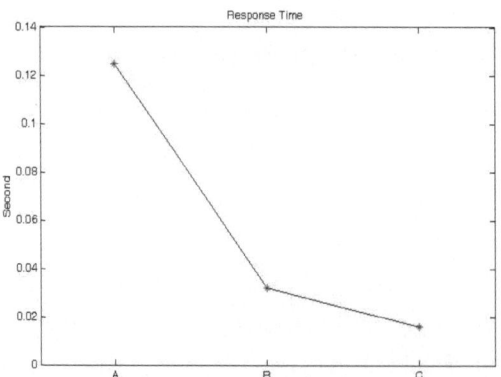

Fig. 7. The comparison of executing time

The improvement of soft set approach for traditional and rough maximal association rules approaches are, 87% and 50%, respectively.

Table 1. The response time improvement of traditional and rough set approaches by soft set approach

	Response Time Improvement
A	87%
B	50%

4 Conclusion

In this paper, we have presented an alternative approach for maximal association rules mining from a transactional database using soft set theory. This approach is started by a transformation of a transactional database into a soft set. Using the co-occurrence of parameters concept in a transaction, we define the notion of a maximal association rule between two maximal itemsets (parameters), its maximal support and maximal confidence under soft set theory. We elaborate the proposed approach through a benchmark database for text categorization and a database of air pollution in Kuala Lumpur on July 2002. We have shown that the maximal rules discovered with the min$M$sup and min$M$cof thresholds are identical to the traditional maximal and rough maximal association rules approaches and achieve faster executing time. With this approach, we believe that some applications using soft set theory for mining various levels of association rules and decision support systems through this view will be applicable and easier.

Acknowledgement

This work was supported by the FRGS under the Grant No. Vote 0402, Ministry of Higher Education, Malaysia.

References

[1] Agrawal, R., Imielinski, T., Swami, A.: Mining association rules between sets of items in large databases. In: Proceedings of the ACM SIGMOD International Conference on the Management of Data, pp. 207–216 (1993)

[2] Agrawal, R., Srikant, R.: Fast algorithms for mining association rules. In: Proceedings of the 20th International Conference on Very Large Data Bases (VLDB), pp. 487–499 (1994)

[3] Feldman, R., Aumann, Y., Amir, A., Zilberstein, A., Klosgen, W.: Maximal association rules: a new tool for mining for keywords cooccurrences in document collections. In: Proceedings of the KDD 1997, pp. 167–170 (1997)

[4] Feldman, R., Fresko, M., Kinar, Y., Lindell, Y., Liphstat, O., Rajman, M., Schler, Y., Zamir, O.: Text mining at the term level. In: Żytkow, J.M. (ed.) PKDD 1998. LNCS, vol. 1510, pp. 65–73. Springer, Heidelberg (1998)

[5] Amir, A., Aumann, Y., Feldman, R., Fresco, M.: Maximal Association Rules: A Tool for Mining Associations in Text. Journal of Intelligent Information Systems 25(3), 333–345 (2005)

[6] Guan, J.W., Bell, D.A., Liu, D.Y.: The Rough Set Approach to Association Rule Mining. In: Proceedings of the Third IEEE International Conference on Data Mining, ICDM 2003 (2003)

[7] Bi, Y., Anderson, T., McClean, S.: A rough set model with ontologies for discovering maximal association rules in document collections. Knowledge-Based Systems 16, 243–251 (2003)

[8] Guan, J.W., Bell, D.A., Liu, D.Y.: Mining Association Rules with Rough Sets. In: Studies in Computational Intelligence, pp. 163–184. Springer, Heidelberg (2005)

[9] Pawlak, Z.: Rough sets. International Journal of Computer and Information Science 11, 341–356 (1982)

[10] Pawlak, Z.: Rough sets: A theoretical aspect of reasoning about data. Kluwer Academic Publisher, Dordrecht (1991)

[11] Pawlak, Z., Skowron, A.: Rudiments of rough sets. Information Sciences 177(1), 3–27 (2007)

[12] Molodtsov, D.: Soft set theory-first results. Computers and Mathematics with Applications 37, 19–31 (1999)

[13] Herawan, T., Mat Deris, M.: A direct proof of every rough set is a soft set. In: Proceeding of International Conference AMS 2009 (2009)

[14] Reuters-21578 (2002), http://www.research.att.com/lewis/reuters21578.html

[15] Mat Deris, M., Nabila, N.F., Evans, D.J., Saman, M.Y., Mamat, A.: Association rules on significant rare data using second support. International Journal of Computer Mathematics 83(1), 69–80 (2006)

Soft Set Theoretic Approach for Dimensionality Reduction

Tutut Herawan[1,3], Ahmad Nazari Mohd. Rose[2], and Mustafa Mat Deris[1]

[1] FTMM, Universiti Tun Hussein Onn Malaysia, Johor, Malaysia
[2] FIT, Universiti Darul Iman Malaysia, Terengganu, Malaysia
[3] CIRNOV, Universitas Ahmad Dahlan, Yogyakarta, Indonesia
tutut81@uad.ac.id, anm@udm.edu.my, mmustafa@uthm.edu.my

Abstract. A reduct is a subset of attributes that are jointly sufficient and individually necessary for preserving a particular property of a given information system. The existing reduct approaches under soft set theory are still based on Boolean-valued information system. However, in the real applications, the data usually contain non-Boolean valued. In this paper, an alternative approach for attribute reduction in multi-valued information system under soft set theory is presented. Based on the notion of multi-soft sets and AND operation, attribute reduction can be defined. It is shown that the reducts obtained are equivalent with Pawlak's rough reduction.

Keywords: Information system; Reduct; Soft set theory.

1 Introduction

In many data analysis applications, information and knowledge are stored and represented in an information table, where a set of objects is described by a set of attributes. To this, one practical problem is faced: for a particular property, whether all the attributes in the attribute set are always necessary to preserve this property [1]. Using the entire attribute set for describing the property is time-consuming, and the constructed rules may be difficult to understand, to apply or to verify. In order to deal with this problem, attribute reduction is required. The objective of reduction is to reduce the number of attributes, and at the same time, preserve the property of information. The theory of soft set [2] proposed, by Molodtsov 1999 is a new method for handling uncertain data. Soft sets are called (binary, basic, elementary) neighborhood systems [3]. The soft set is a mapping from parameter to the crisp subset of universe. From such case, we may see the structure of a soft set can classify the objects into two classes (yes/1 or no/0). This means that the "standard" soft set deals with a Boolean-valued information system. The theory of soft set has been applied to data analysis and decision support systems. A fundamental notion supporting such applications is the concept of reducts. The idea of dimensionality reduction under soft set theory have been proposed and compared, including the works of [4-8]. The restriction of those techniques is that they are applicable only for Boolean-valued information systems. However, in the theoretical and practical

D. Ślęzak et al. (Eds.): DTA 2009, CCIS 64, pp. 171–178, 2009.
© Springer-Verlag Berlin Heidelberg 2009

researches of soft sets, the situations are usually very complex. In the real application, depending on the set of parameters, a given parameter may have different values (contain multiple grades). For example, the mathematics degree of student can be classified into three values; high, medium and low. In this situation, every parameter determines a partition of the universe which is contains more than two disjoint subsets. Unlike in Boolean-valued information systems, in multi-valued information systems, one cannot directly define the standard soft set. To this, we proposed the idea of *multi-soft sets* to deal multi-valued information systems. In this paper, we propose the idea of dimensionality reduction for multi-valued information systems under soft set theory. Three main contributions of this work are as follows: Firstly, we present the idea of multi-soft sets construction from a multi-valued information system, and AND and OR operations on multi-soft sets. Secondly, we present the applicability of the soft set theory for data reduction under multi-valued information system using multi-sets and AND operation. Lastly, we show that reducts obtained using soft set theory are identical to that rough set theory. Although some results are presented, a major part of this paper is devoted to revealing interconnections between reduction in multi-valued information systems under rough and soft set theories.

The rest of this paper is organized as follows. Section 2 describes related works of dimensionality reduction under soft set theory. Section 3 describes Information systems and set approximations. Section 4 describes the definition of soft set. Section 5 describes reduct in information systems using soft set theory. Finally, we conclude our works in section 6.

2   Related Works

The idea of reduct and decision making using soft set theory is firstly proposed by Maji *et al.* [4]. In [4], the application of soft set theory to a decision making problem with the help of Pawlak's rough mathematics is presented. The reduction approach presented is using Pawlak's rough reduction and a decision can be selected based on the maximal weighted value among objects related to the parameters. Chen *et al.* [5-6] presented the parameterization reduction of soft sets and its applications. They pointed out that the results of reduction proposed by Maji is incorrect and observed that the algorithms used to compute the soft set reduction and then to compute the choice value to select the optimal objects for the decision problem proposed by Maji are unreasonable. They also pointed out that the idea of reduct under rough set theory generally cannot be applied directly in reduct under soft set theory. The idea of Chen for soft set reduction is only based on the optimal choice related to each object. However, the idea proposed by Chen is not error free, since the problems of the sub-optimal choice is not addressed. To this, Kong *et al.* [7] analyzed the problem of suboptimal choice and added parameter set of soft set. Then, they introduced the definition of normal parameter reduction in soft set theory to overcome the problems in Chen's model and described two new definitions, i.e. parameter important degree and soft decision partition and use them to analyze the algorithm of normal parameter reduction. With this approach, the optimal and sub-optimal choices are still preserved. Zou [8] proposed a new technique for decision making of soft set theory under incomplete information systems. The idea is based on the calculation of weighted-average of all possible choice values of object and the weight of each possible choice value is decided by the distribution of other objects. For fuzzy soft sets, incomplete data will be predicted based on the method of average probability. All those techniques are still based on Boolean-valued information systems. No researches have been done on dimensionality reduction in multi-valued information systems under soft set theory. Since every rough set [9] can be considered as soft set [10], thus, an alternative approach with potential for finding reduct in multi-valued information systems is using soft set theory. Still, it provides the same results for rough reduction [11-12].

3 Information Systems and Set Approximations

The definition of an information system $S = (U, A, V, f)$ is the same as in [12]. The starting point of rough set approximations is the indiscernibility (equivalence) relation, which is generated by information about objects of interest.

Definition 1. (See [12].) *Let $S = (U, A, V, f)$ be an information system and let B be any subset of A. Two elements $x, y \in U$ are said to be B-indiscernible (indiscernible by the set of attribute $B \subseteq A$ in S) if and only if $f(x, a) = f(y, a)$, for every $a \in B$.*

The partition of U induced by B is denoted by U / B and the equivalence class in the partition U / B containing $x \in U$, denoted by $[x]_B$. The notions of lower and upper approximations of a set can be defined as follows.

Definition 2. (See [12].) *Let $S = (U, A, V, f)$ be an information system, let B be any subset of A and let X be any subset of U. The B-lower approximation of X, denoted by $\underline{B}(X)$ and B-upper approximation of X, denoted by $\overline{B}(X)$, respectively, are defined by $\underline{B}(X) = \{x \in U \mid [x]_B \subseteq X\}$ and $\overline{B}(X) = \{x \in U \mid [x]_B \cap X \neq \phi\}$.*

The notions of rough approximating of a set can be defined as follows:

Definition 3. *Let $S = (U, A, V, f)$ be an information system and let B be any subset of A. A rough approximation of a subset $X \subseteq U$ with respect to B is defined as a pair of lower and upper approximations of X, i.e. $\langle \underline{B}(X), \overline{B}(X) \rangle$.*

Definition 4. *Let $S = (U, A, V, f)$ be an information system and let B be any subsets of A. Attribute $b \in B$ is called dispensable if*

$$U / (B - \{b\}) = U / B.$$

Definition 5. *Let $S = (U, A, V, f)$ be an information system and let B be any subset of A. The subset $B^* \subseteq B$ is called reduct of B if B^* satisfies $U / B^* = U / B$ and*

$U/(B*-\{b\}) \neq U/B$, $\forall b \in B*$. The core of B is defined as $\text{CORE}(B) = \bigcap \text{RED}(B)$, where $\text{RED}(B)$ is the set of all reducts of B.

It is known that the problem of finding minimal reducts in information systems is NP-hard.

Example 6. For simple example of rough reduction we consider a small dataset derived from [13].

Table 1. An information system from [13]

U	a_1	a_2	a_3	a_4
1	low	bad	loss	small
2	low	good	loss	large
3	high	good	loss	medium
4	high	good	loss	medium
5	low	good	profit	large

Let $A = \{a_1, a_2, a_3, a_4\}$, then we have $U/A = \{\{1\}, \{2\}, \{3,4\}, \{5\}\}$. Since $U/B = U/C = \{\{1\}, \{2\}, \{3,4\}, \{5\}\} = U/A$, then the reducts of A are $B = \{a_1, a_2, a_3\}$ and $C = \{a_3, a_4\}$ are reducts of B. The core is $\text{CORE}(A) = B \cap C = \{a_3\}$.

The definition of soft set and its relations with Boolean-valued information system and rough set are given in the following section. The definitions are quoted directly from [2,14].

4   Soft Set Theory

Throughout this section U refers to an initial universe, E is a set of parameters, $P(U)$ is the power set of U and $A \subseteq E$.

Definition 7. (See [2].) *A pair* (F, A) *is called a soft set over U, where F is a mapping given by* $F : A \rightarrow P(U)$.

In other words, a soft set over U is a parameterized family of subsets of the universe U. For $\varepsilon \in A$, $F(\varepsilon)$ may be considered as the set of ε-elements of the soft set (F, A) or as the set of ε-approximate elements of the soft set. Clearly, a soft set is not a (crisp) set.

Definition 8. (See [14].) *The class of all value sets of a soft set* (F, E) *is called value-class of the soft set and is denoted by* $C_{(F,E)}$.

Clearly $C_{(F,E)} \subseteq P(U)$.

Proposition 9. *If* (F, E) *is a soft set over the universe U, then* (F, E) *is a Boolean-valued information system* $S = (U, A, V, f)$.

Proof. Let (F, E) be a soft set over the universe U, we define a collection of a mappings $F = \{f_1, f_2, \cdots, f_n\}$, where

$$f_i : U \to V_i \text{ and } f_i(x) = \begin{cases} 1, & x \in F(e_i) \\ 0, & x \notin F(e_i) \end{cases}, \text{ for } i = 1, \cdots, n.$$

Thus, if $A = E$, $V = \bigcup_{e_i \in A} V_{e_i}$, where $V_{e_i} = \{0,1\}$, then a soft set (F, E) can be considered as a Boolean-valued information system $S = (U, A, V_{\{0,1\}}, f)$. □

From Proposition 10, it is easily to understand that a Boolean-valued information system can be represented as a soft set. Thus, we can make a one-to-one correspondence between (F, E) over U and $S = (U, A, V_{\{0,1\}}, f)$.

Proposition 10. *Every rough set can be considered as a soft set.*

Let $\langle \underline{B}(X), \overline{B}(X) \rangle$ be the rough approximating a subset $X \subseteq U$. We define a mapping $\underline{B}, \overline{B} : P(U) \to P(U)$ as in Definition 3. Thus every rough set $\langle \underline{B}(X), \overline{B}(X) \rangle$ can be considered a pair of two soft sets $(F, U) = \langle (\underline{B}, P(U)), (\overline{B}, P(U)) \rangle$. □

From the fact that every rough set can be considered as a soft set, in the following section we propose an alternative approach for dimensionality reduction in multi-valued information systems under soft set theory.

5 Reduction in Information Systems Using Soft Set Theory

In this section we present the applicability of soft set theory for finding reducts. We show that the reducts obtained are similar to the rough reducts as in [11-12]. The first step, we need a transformation from a multi-valued information system into *multi-soft sets*. In the multi-soft sets, we present the notion of AND and OR operations. For attribute reduction, we employ AND operation and show that the reducts obtained are equivalent to rough reducts.

5.1 Multi-soft Sets, AND and OR Operations

The notion of multi-soft sets has been presented in [15]. The idea is based on a decomposition of a multi-valued information system $S = (U, A, V, f)$ into $|A|$ number of Boolean-valued information systems.

Example 11. The multi-soft sets representing Table 1 is given by

$$(F,A) = ((F,a_1),(F,a_2),(F,a_3),(F,a_4))$$
$$= \begin{pmatrix} \{\{low = 1,2,5\}, \{high = 3,4\}\}, \{\{bad = 1\}, \{good = 2,3,4,5\}\}, \\ \{\{loss = 1,2,3,4\}, \{profit = 5\}\}, \{\{small = 1\}, \{large = 3,4\}, \{medium = 3,4\}\} \end{pmatrix}$$

The notions of AND and OR operations in multi-soft sets are given below.

Definition 12. *Let* $(F,E) = ((F,a_i): i = 1,2,\cdots,|A|)$ *be a multi-soft set over* U *representing a multi-valued information system* $S = (U,A,V,f)$ *. The AND operation between* (F,a_i) *and* (F,a_j) *is defined as* $(F,a_i)AND(F,a_j) = (F,a_i \times a_j)$*, where* $F(Va_i,Va_j) = F(Va_i) \cap F(Va_j)$*,* $\forall (Va_i,Va_j) \in a_i \times a_j$*, for* $1 \le i, j \le |A|$*.*

Example 13. From Example 11, let two soft-sets

$$(F,a_1) = \{\{low = 1,2,5\}, \{high = 3,4\}\} \text{ and } (F,a_2) = \{\{bad = 1\}, \{good = 2,3,4,5\}\}.$$
Then, we have $(F,a_1)AND(F,a_2) = (F,a_1 \times a_2)$
$$= ((low, bad) = \{1\}, (low, good) = \{2,5\}, (high, bad) = \phi, (high, good) = \{3,4\}).$$

Definition 14. *Let* $(F,E) = ((F,a_i): i = 1,2,\cdots,|A|)$ *be a multi-soft set over* U *representing a multi-valued information system* $S = (U,A,V,f)$*. The OR operation between* (F,a_i) *and* (F,a_j) *is defined as* $(F,a_i)OR(F,a_j) = (F,a_i \times a_j)$*, where* $G(Va_i,Va_j) = F(Va_i) \cup F(Va_j)$*,* $\forall (Va_i,Va_j) \in a_i \times a_j$*, for* $1 \le i, j \le |A|$*.*

Example 15. From Example 11, let two soft-sets

$$(F,a_2) = \{\{bad = 1\}, \{good = 2,3,4,5\}\} \text{ and } (F,a_3) = \{\{loss = 1,2,3,4\}, \{profit = 5\}\},$$
Then we have $(F,a_2)OR(F,a_3) = (F,a_2 \times a_3) =$
$$= \begin{pmatrix} (bad, loss) = \{1,2,3,4\}, (bad, profit) = \{1,5\}, \\ (good, loss) = \{1,2,3,4,5\}, (good, profit) = \{2,3,4,5\} \end{pmatrix}$$

5.2 Attribute Reduction

In this section we propose the idea of attributes reduction under soft set theory. The proposed approach is based on AND operation in multi-soft sets as described in the previous section.

Definition 16. *Let* $(F,A) = ((F,a_i): i = 1,2,\cdots,|A|)$ *be a multi-soft set over* U *representing a multi-valued information system* $S = (U,A,V,f)$*. A set of attributes* $B \subseteq A$ *is called a reduct for A if* $C_{F(b_1 \times \cdots \times b_{|B|})} = C_{F(a_1 \times \cdots \times a_{|A|})}$*.*

Example 17. From Example 11, let two multi soft-sets

$$(F,\{a_1,a_2,a_3\}) \text{ and } (F,\{a_3,a_4\})$$

a.   For $(F\{a_1,a_2,a_3\})$, where $(F,a_1) = \{\{\text{low} = 1,2,5\}, \{\text{high} = 3,4\}\}$, $(F,a_2) = \{\{\text{bad} = 1\}, \{\text{good} = 2,3,4,5\}\}$ and $(F,a_3) = \{\{\text{loss} = 1,2,3,4\},\{\text{profit} = 5\}\}$.

Then, we have

$$(F,a_1)\text{AND}(F,a_2)\text{AND}(F,a_3) = (F,a_1 \times a_2 \times a_3)$$

$$= \begin{pmatrix} (\text{low, bad, loss}) = \{1\}, (\text{low, bad, profit}) = \{5\}, \\ (\text{low, good, loss}) = \{2\}, (\text{low, good, profit}) = \{5\}, \\ (\text{high, bad, loss}) = \phi, (\text{high, bad, profit}) = \phi, \\ (\text{high, good, loss}) = \{3,4\}, (\text{high, good, profit}) = \{5\} \end{pmatrix}$$

Notice that,

$$C_{F(a_1 \times a_2 \times a_3)} = \{\{1\},\{2\},\{3,4\},\{5\}\}. \tag{1}$$

b.   For $(F,\{a_3,a_4\})$, where $(F,a_3) = \{\{\text{loss} = 1,2,3,4\},\{\text{profit} = 5\}\}$ and $(F,a_4) = \{\{\text{small} = 1\}, \{\text{large} = 2,5\}, \{\text{medium} = 3,4\}\}$,

Then we have

$$(F,a_3)\text{AND}(F,a_4) = (F,a_3 \times a_4)$$

$$= \begin{pmatrix} (\text{loss, small}) = \{1\}, (\text{loss, large}) = \{2\}, (\text{loss, medium}) = \{3,4\}, \\ (\text{profit, small}) = \phi, (\text{profit, large}) = \{5\}, (\text{profit, medium}) = \phi \end{pmatrix}$$

Notice that,

$$C_{F(a_3 \times a_4)} = \{\{1\},\{2\},\{3,4\},\{5\}\}. \tag{2}$$

From (1) and (2), we have $\{a_1,a_2,a_3\}$ and $\{a_3,a_4\}$ are reducts of A.

6   Conclusion

The existing reduct approaches under soft set theory are still based on Boolean-valued information system. For the real applications, the data usually contain non-Boolean valued. In this paper, an alternative approach for attribute reductions in multi-valued information systems under soft set theory using soft set theory has been presented. In the proposed approach, the notion of multi-soft set is used to represent multi-valued information systems. The AND operation is used in multi-soft sets to present the notion of attribute reduction. It is founded that the obtained reducts are equivalent to the rough reducts. In the next papers, we plan to apply the presented method to design algorithms for solving the data cleansing in larger data sets, decision making, data clustering problems.

Acknowledgement

This work was supported by the FRGS under the Grant No. Vote 0402, Ministry of Higher Education, Malaysia.

References

[1] Zhao, Y., Luo, F., Wong, S.K.M., Yao, Y.Y.: A general definition of an attribute reduct. In: Yao, J., Lingras, P., Wu, W.-Z., Szczuka, M.S., Cercone, N.J., Ślęzak, D. (eds.) RSKT 2007. LNCS (LNAI), vol. 4481, pp. 101–108. Springer, Heidelberg (2007)

[2] Molodtsov, D.: Soft set theory-first results. Computers and Mathematics with Applications 37, 19–31 (1999)

[3] Yao, Y.Y.: Relational interpretations of neighbourhood operators and rough set approximation operators. Information Sciences 111, 239–259 (1998)

[4] Maji, P.K., Roy, A.R., Biswas, R.: An application of soft sets in a decision making problem. Computer and Mathematics with Application 44 (2002)

[5] Chen, D., Tsang, E.C.C., Yeung, D.S., Wang, X.: Some notes on the parameterization reduction of soft sets. In: Proceeding of International Conference on Machine Learning and Cybernetics, vol. 3, pp. 1442–1445 (2003)

[6] Chen, D., Tsang, E.C.C., Yeung, D.S., Wang, X.: The Parameterization Reduction of Soft Sets and its Applications. Computers and Mathematics with Applications 49, 757–763 (2005)

[7] Zhi, K., Gao, L., Wang, L., Li, S.: The normal parameter reduction of soft sets and its algorithm. Computers and Mathematics with Applications 56, 3029–3037 (2008)

[8] Zou, Y., Xiao, Z.: Data analysis approaches of soft sets under incomplete information. Knowledge Based Systems 21, 941–945 (2008)

[9] Pawlak, Z.: Rough sets. International Journal of Computer and Information Science 11, 341–356 (1982)

[10] Herawan, T., Mat Deris, M.: A direct proof of every rough set is a soft set. In: Proceeding of IEEE International Conference AMS 2009, pp. 119–124 (2009)

[11] Pawlak, Z.: Rough sets: A theoretical aspect of reasoning about data. Kluwer Academic Publishers, Dordrecht (1991)

[12] Pawlak, Z., Skowron, A.: Rudiments of rough sets. Information Sciences. An International Journal 177(1), 3–27 (2007)

[13] Pawlak, Z.: Rough classification. International journal of human computer studies 51, 369–383 (1999)

[14] Maji, P.K., Biswas, R., Roy, A.R.: Soft set theory. Computers and Mathematics with Applications 45, 555–562 (2003)

[15] Herawan, T., Mat Deris, M.: On multi-soft sets construction in information systems. In: Manuscript accepted at ICIC 2009 (2009) (to appear in LNAI Springer)

Rough Set Approach for Categorical Data Clustering

Tutut Herawan[1,2], Iwan Tri Riyadi Yanto[1], and Mustafa Mat Deris[1]

[1] FTMM, Universiti Tun Hussein Onn Malaysia, Johor, Malaysia
[2] CIRNOV, Universitas Ahmad Dahlan, Yogyakarta, Indonesia
tutut81@uad.ac.id, iwan015@gmail.com, mmustafa@uthm.edu.my

Abstract. In this paper, we focus our discussion on the rough set approach for categorical data clustering. We propose MADE (Maximal Attributes Dependency), an alternative technique for categorical data clustering using rough set theory taking into account maximal attributes dependencies. Experimental results on two benchmark UCI datasets show that MADE technique is better with the baseline categorical data clustering techniques with respect to computational complexity and clusters purity.

Keywords: Clustering, Categorical data, Rough set theory, Attributes dependencies.

1 Introduction

Clustering a set of objects into homogeneous classes is a fundamental operation in data mining. Recently, many attentions have been put on categorical data clustering [1,2], where data objects are made up of non-numerical attributes. One of the popular approaches is based on rough set theory [3]. The main idea of the rough clustering is the clustering database is mapped as the decision table. The first attempt on rough set-based technique to select clustering attribute is proposed by Mazlack *et al.* [4]. They proposed two techniques, i.e., Bi-Clustering and TR techniques. One of the most successful pioneering rough clustering techniques is Minimum-Minimum Roughness (MMR) proposed by Parmar [5]. However, since application of rough set theory in categorical data clustering is relatively new, the focus of MMR is still on evaluating its performance. To this, the computational complexity and clusters purity are still an issue due to all attributes are considered to be selected and objects in different class appear in one clusters, respectively. In this paper, we propose MADE (Maximal Attributes Dependency), an alternative technique for categorical data clustering. The technique differs on the baseline techniques and the rough attributes dependency in a categorical-valued information system is used to select clustering attribute based on the maximal degree. Further, we use a divide-and-conquer method to partition/cluster the objects. We show that the proposed method is a generalization of MMR, achieve lower computational complexity with higher purity compared to MMR. The rest of this paper is organized as follows. Section 2 describes the concept of rough set theory in an information system, the notion of attributes dependency, TR and MMR techniques. Section 3 describes the Maximum Attributes Dependency (MADE)

D. Ślęzak et al. (Eds.): DTA 2009, CCIS 64, pp. 179–186, 2009.
© Springer-Verlag Berlin Heidelberg 2009

technique. Comparison tests of MADE with MMR techniques based on Soybean and Zoo databases are described in section 4. Finally, the conclusion of this work is described in section 5.

2   Rough Set Approach for Selecting Clustering Attribute

2.1   Rough Set Theory

The syntax of information systems is very similar to relations in relational data bases. Entities in relational databases are also represented by tuples of attribute values. An *information system* as in [6] is a 4-tuple (quadruple) $S = (U, A, V, f)$, where $U = \{u_1, u_2, u_3, \cdots, u_{|U|}\}$ is a non-empty finite set of objects, $A = \{a_1, a_2, a_3, \cdots, a_{|A|}\}$ is a non-empty finite set of attributes, $V = \bigcup_{a \in A} V_a$, V_a is the domain (value set) of attribute a, $f : U \times A \rightarrow V$ is an information function such that $f(u, a) \in V_a$, for every $(u, a) \in U \times A$, called information (knowledge) function. An information system is also called a knowledge representation systems or an attribute-valued system and can be intuitively expressed in terms of an information table (see Table 1).

Table 1. An information system

U	a_1	a_2	$\cdots$	a_k	$\cdots$	$a_{|A|}$
u_1	$f(u_1, a_1)$	$f(u_1, a_2)$	$\cdots$	$f(u_1, a_k)$	$\cdots$	$f(u_1, a_{|A|})$
u_2	$f(u_2, a_1)$	$f(u_2, a_2)$	$\cdots$	$f(u_2, a_k)$	$\cdots$	$f(u_2, a_{|A|})$
$\vdots$	$\vdots$	$\vdots$	$\ddots$	$\vdots$	$\ddots$	$\vdots$
$u_{|U|}$	$f(u_{|U|}, a_1)$	$f(u_{|U|}, a_2)$	$\cdots$	$f(u_{|U|}, a_k)$	$\cdots$	$f(u_{|U|}, a_{|A|})$

Two elements $x, y \in U$ are said to be *B-indiscernible* (indiscernible by the set of attribute $B \subseteq A$ in S) if and only if $f(x, a) = f(y, a)$, for every $a \in B$. The partition of U induced by B is denoted by U/B and the equivalence class in the partition U/B containing $x \in U$, is denoted by $[x]_B$. The notions of lower and upper approximations of a set are defined as follows. The *B-lower approximation* of $X \subseteq U$, denoted by $\underline{B}(X)$ and *B-upper approximation* of X, denoted by $\overline{B}(X)$ respectively, are defined by

$$\underline{B}(X) = \{x \in U \mid [x]_B \subseteq X\} \text{ and } \overline{B}(X) = \{x \in U \mid [x]_B \cap X \neq \phi\}.$$

The accuracy of approximation of any subset $X \subseteq U$ with respect to $B \subseteq A$, denoted $\alpha_B(X)$ is measured by $\alpha_B(X) = |\underline{B}(X)| / |\overline{B}(X)|$, where $|X|$ denotes the cardinality of X. If $\alpha_B(X) = 1$, the set X is said to be *crisp*, and otherwise, X is *rough* with respect to B. The notion of attributes dependency is given in the following definition.

Definition 1. *Let* $S = (U, A, V, f)$ *be an information system and let D and C be any subsets of A. The degree of dependency of attribute D on attributes C, denoted* $C \Rightarrow_k D$, *is defined by*

$$k = \sum_{X \in U/D} |\underline{C}(X)| / |U|.$$

Attribute *D* is said to be depends totally on the attribute *C* if $k = 1$. Otherwise, *D* is depends partially on *C*.

2.2 TR and MMR Techniques

The analysis and comparison of TR and MMR Techniques has been presented in [7]. We show that TR and MMR techniques produce the same results in terms of determining the clustering attribute and complexity. Furthermore, to achieve lower computational complexity in selecting clustering attribute using MMR, Parmar *et al.* suggested to measure the roughness based on relationship between an attribute $a_i \in A$ and the set defined as $A - \{a_i\}$ instead of calculating the maximum with respect to all $\{a_j\}$ where $a_i \neq a_j$. We observe this technique only can be applied in a special database.

We let the dataset in illustrative example of Table 2 in [5] to explain this problem. If we measure the roughness of attribute $a_i \in A$ with respect to the set of attributes $A - \{a_i\}$, then we get the value of modified MMR is 0, for all attributes a_i with respect to attribute $A - \{a_i\}$. Thus, we cannot select a clustering attribute. Therefore, the suggested technique would lead a problem, i.e., after calculation of mean roughness of attribute $a_i \in A$ with respect to the set of attributes $A - \{a_i\}$, the value of MMR usually cannot preserve the original decision. Hence, this modified technique is not relevant to all type of data set.

To overcome the problem of computational complexity of MMR, in the next section, we introduce the Maximum Attributes Dependencies (MADE) technique.

3   Maximum Attributes Dependencies (MADE) Technique

3.1 MADE Technique

The MADE technique for selecting partitioning attribute is based on the maximum degree of dependency of attributes. The justification that the higher of the degree of dependency of attributes implies the more accurate for selecting partitioning attribute is stated in the Proposition 2.

Proposition 2. *Let* $S = (U, A, V, f)$ *be an information system and let D and C be any subsets of A. If D depends totally on C, then*

$$\alpha_D(X) \leq \alpha_C(X),$$

for every $X \subseteq U$.

Proof. Let D and C be any subsets of A in information system $S = (U, A, V, f)$. From the hypothesis, we have $IND(C) \subseteq IND(D)$. Furthermore, the partitioning U/C is finer than that U/D, thus, it is clear that any equivalence class induced by $IND(D)$ is a union of some equivalence class induced by $IND(C)$. Therefore, for every $x \in X \subseteq U$, we have $[x]_C \subseteq [x]_D$. And hence, for every $X \subseteq U$, we have

$$\underline{D}(X) \subseteq \underline{C}(X) \subset X \subset \overline{C}(X) \subseteq \overline{D}(X).$$

Consequently

$$\alpha_D(X) = \underline{D}(X) / |\overline{D}(X)| \le |\underline{C}(X)| / |\overline{C}(X)| = \alpha_C(X). \qquad \square$$

3.2 Complexity

Suppose that in an information system $S = (U, A, V, f)$, there is $|A|$ attributes. For MADE, the computation of calculating of dependency degree of attribute a_i on attribute a_j, where $i \ne j$ is $|A| \times |A-1|$. Thus, the computational complexity for MADE technique is of the polynomial $O(|A| \times |A-1|)$.

The MADE's pseudo-code for selecting clustering attribute is given in Figure 1.

MADE Pseudo-code
Input: A database without decision attribute Output: Decision attribute

``
function R=MADE(A); A the matrix of a database
[m n]=size(A);
while it<n
    it=it+1;
    itc=0;acc=[];
    while itc<n
        itc=itc+1;
        if itc~=it
            itA=A(:,it);
            itB=A(:,itc);
            %dependency
             smade=feval('depend',itA,itB);
            acc1=smade./m;
            acc=[acc acc1];
        end
    end
     acc3=[acc3; acc];
end
``

Fig. 1. MADE's pseudo-code

As the same procedure for selecting clustering attribute of MMR, in using MADE technique, it is recommended to look at the next lowest dependencies degree inside the attributes that are tied and so on until the tie is broken.

3.3 Objects Splitting

For objects splitting, we use a divide-conquer method. The technique is applied recursively to obtain further clusters. At subsequent iterations, the leaf node having more objects is selected for further splitting. The technique terminates when it reaches a pre-defined number of clusters. This is subjective and is pre-decided based either on user requirement or domain knowledge.

4 Comparison Tests

In order to test MADE and compare with MMR, we use two databases obtained from the benchmark UCI Machine Learning Repository. We use Soybean and Zoo databases are with 47 and 101 objects. The purity of clusters was used as a measure to test the quality of the clusters [5]. The purity of a cluster and overall purity are defined as

$$\text{Purity}(i) = \frac{\text{the number of data occuring in both the } i\text{th cluster and its corresponding class}}{\text{the number of data in the data set}}$$

$$\text{Overall Purity} = \frac{\sum_{i=1}^{\#\text{ of cluster}} \text{Purity}(i)}{\#\text{ of cluster}}$$

According to this measure, a higher value of overall purity indicates a better clustering result, with perfect clustering yielding a value of 1 [5]. The techniques of MMR and MADE for Soybean and Zoo databases are implemented in MATLAB version 7.6.0.324 (R2008a). They are executed sequentially on a processor Intel Core 2 Duo CPUs. The total main memory is 1 Gigabyte and the operating system is Windows XP Professional SP3.

4.1 Soybean Database

The Soybean database contains 47 objects on diseases in soybeans. Each object can be classified as one of the four diseases namely, Diaporthe Stem Canker (D1), Charcoal Rot (D2), Rhizoctonia Root Rot (D3), and Phytophthora Rot (D4) and are described by 35 categorical attributes [8]. The database is comprised 17 objects for Phytophthora Rot disease and 10 objects for each of the remaining diseases. Since there are four possible diseases, the objects will be split into four clusters. The results are summarized in Table 2. All of 47 objects belong to the majority class label of the cluster in which they are classified. Thus, the overall purity of the clusters is 100%.

Table 2. The purity of clusters

Cluster number	D1	D2	D3	D4	Purity
1	10	0	0	0	1
2	0	10	0	0	1
3	0	0	10	0	1
4	0	0	0	17	1
Overall Purity					1

4.2 Zoo Database

The Zoo database is comprised of 101 objects, where each data point represents information of an animal in terms of 18 categorical attributes [9]. Each animal data point is classified into seven classes. Therefore, for MADE, the splitting data is set at seven clusters. Table 3 summarizes the results of running the MADE technique on the Zoo database.

Table 3. The purity of clusters

Cluster number	C1	C2	C3	C4	C5	C6	C7	Purity
1	41	0	0	0	0	0	0	1
2	0	20	0	0	0	0	0 ·	1
3	0	0	5	0	0	0	0	1
4	0	0	0	13	0	0	0	1
5	0	0	0	0	4	0	0	1
6	0	0	0	0	0	8	0	1
7	0	0	0	0	0	0	10	1
Overall Purity								1

All of 101 objects belong to the majority class label of the cluster in which they are classified. Thus, the overall purity of the clusters is 100%.

4.3 Comparison Results

The comparison of overall purity, computation and response time of MADE and MMR on Soybean and Zoo databases are given in Figures 2, 3 and 4, respectively. Based on Table 4, the MADE technique provides better solution compared to MMR technique both in Soybean and Zoo database.

Table 4. The overall improvement of MMR by MDA

	Improvement		
	Clusters Purity	Computation	Response Time
Soybean	17%	64%	63%
Zoo	9%	77%	67%

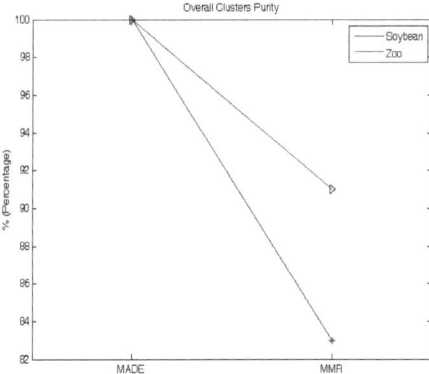

Fig. 2. The comparison of overall purity

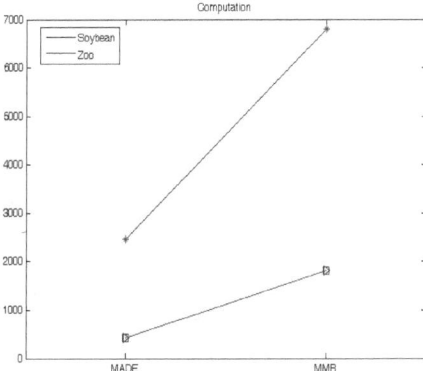

Fig. 3. The computation

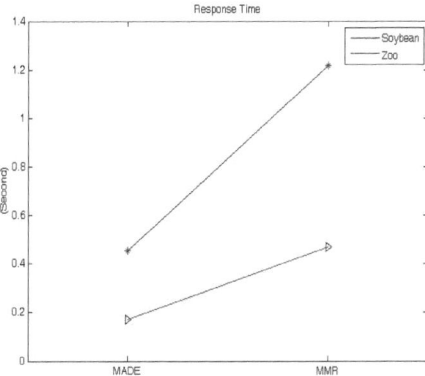

Fig. 4. The response time

5 Conclusion

In this paper, we have proposed MADE, an alternative technique for categorical data clustering using rough set theory based on attributes dependencies. We have shown that MADE technique provides lower computational complexity and higher clusters purity as compared to MMR. With this approach, we believe that some applications through MADE will be applicable, such as for decision making and etc.

Acknowledgement

This work was supported by the FRGS under the Grant No. Vote 0402, Ministry of Higher Education, Malaysia.

References

[1] Huang, Z.: Extensions to the k-means algorithm for clustering large data sets with categorical values. Data Mining and Knowledge Discovery 2(3), 283–304 (1998)

[2] Kim, D., Lee, K., Lee, D.: Fuzzy clustering of categorical data using fuzzy centroids. Pattern Recognition Letters 25(11), 1263–1271 (2004)

[3] Pawlak, Z.: Rough sets. International Journal of Computer and Information Science 11, 341–356 (1982)

[4] Mazlack, L.J., He, A., Zhu, Y., Coppock, S.: A rough set approach in choosing partitioning attributes. In: Proceedings of the ISCA 13th, International Conference, CAINE 2000, pp. 1–6 (2000)

[5] Parmar, D., Wu, T., Blackhurst, J.: MMR: An algorithm for clustering categorical data using rough set theory. Data and Knowledge Engineering 63, 879–893 (2007)

[6] Pawlak, Z., Skowron, A.: Rudiments of rough sets. International Journal Information Sciences 177(1), 3–27 (2007)

[7] Herawan, T., Mat Deris, M.: Rough set theory for selecting clustering attribute. In: Manuscript accepted at PCO 2009, Bali Indonesia (2009) (to appear in AIP)

[8] http://archive.ics.uci.edu/ml/datasets/Soybean+%28Small%29

[9] http://archive.ics.uci.edu/ml/datasets/Zoo

Author Index